Lecture Notes in Computer Science 13862

The series Lecture Notes in Computer Science (LNCS), including its subseries Lecture Notes in Artificial Intelligence (LNAI) and Lecture Notes in Bioinformatics (LNBI), has established itself as a medium for the publication of new developments in computer science and information technology research, teaching, and education.

LNCS enjoys close cooperation with the computer science R & D community, the series counts many renowned academics among its volume editors and paper authors, and collaborates with prestigious societies. Its mission is to serve this international community by providing an invaluable service, mainly focused on the publication of conference and workshop proceedings and postproceedings. LNCS commenced publication in 1973.

Eckhard Hitzer · George Papagiannakis ·
Petr Vasik
Editors

Empowering Novel Geometric Algebra for Graphics and Engineering

7th International Workshop, ENGAGE 2022
Virtual Event, September 12, 2022
Proceedings

Editors
Eckhard Hitzer
International Christian University
Tokyo, Japan

Petr Vasik
Brno University of Technology
Brno, Czech Republic

George Papagiannakis
Department of Computer Science
University of Crete
Heraklion, Greece

ISSN 0302-9743 ISSN 1611-3349 (electronic)
Lecture Notes in Computer Science
ISBN 978-3-031-30922-9 ISBN 978-3-031-30923-6 (eBook)
https://doi.org/10.1007/978-3-031-30923-6

This Springer imprint is published by the registered company Springer Nature Switzerland AG
The registered company address is: Gewerbestrasse 11, 6330 Cham, Switzerland

Preface

Welcome to the Lecture Notes in Computer Science (LNCS) proceedings of the workshop Empowering Novel Geometric Algebra for Graphics & Engineering (ENGAGE 2022) at the Computer Graphics International conference (CGI 2022). CGI is one of the oldest international conferences in Computer Graphics in the world. The ENGAGE workshop has been organized since 2016 as part of the annual CGI conference. CGI is the official conference of the Computer Graphics Society (CGS), a long-standing international computer graphics organization. CGI has been held in many different countries across the world and gained a reputation as one of the key conferences for researchers and practitioners to share their achievements and discover the latest advances in Computer Graphics. In 2022, CGI 2022 (and the ENGAGE 2022 workshop) continued to be virtual due to the pandemic that was still all over the place. ENGAGE 2022 was organized online by MIRALab, University of Geneva. The official date of the workshop was September 12, 2022. Finally, some of the presentations from ENGAGE 2022 are available on YouTube.

In retrospective, the ACM Siggraph 2001 and 2003 conferences saw Geometric Algebra (GA) featured in the form of a Keynote and a Course. Since then, the GA community has highlighted the benefits of employing W. K. Clifford's GA, quaternions and octonions for computer graphics and vision problems. The Siggraph 2019 course on projective GA (PGA), GAME 2020 and the Siggraph 2022 course on GA further boosted GA and associated algebras as a language for graphics. The advances were presented at the Workshops CGI 2016 on "Geometric Algebra in Computer Science and Engineering" and annually at CGI 2017–2022 in the ENGAGE workshops and have underlined the power of GA for analysis, computation and graphics.

This extraordinary ENGAGE 2022 LNCS proceedings (due to special circumstances) is composed of 10 papers. We accepted 10 papers from 12 invited submissions. To ensure the high quality of publications, each paper was reviewed double blind by at least three experts in the field and the authors of accepted papers were asked to thoroughly revise their papers taking into account the review comments prior to publication.

The accepted papers focused specifically on important aspects of geometric algebra including algebraic foundations, digitized transformations, orientation, conic fitting, protein modelling, digital twinning, and multidimensional signal processing.

We would like to express our deepest gratitude to the CGI 2022 organizers for again hosting the ENGAGE workshop, and to all the PC members and external reviewers who provided timely, high-quality reviews. We would also like to thank all the authors for contributing to the workshop by submitting their work.

January 2023

Eckhard Hitzer
George Papagiannakis
Petr Vasik

Organization

Workshop and Program Chairs

Eckhard Hitzer	International Christian University, Japan
George Papagiannakis	University of Crete, Greece
Petr Vasik	Brno University of Technology, Czech Republic

Workshop and Program Committee

Andreas Aristidou	University of Cyprus, Cyprus
Werner Benger	Louisiana State University, USA and Airborne HydroMapping GmbH, Austria
Dietmar Hildenbrand	Technical University of Darmstadt, Germany
Eckhard Hitzer	International Christian University, Japan
Joan Lasenby	University of Cambridge, UK
Kit Ian Kou	University of Macao, China
Vincent Nozick	Université Gustave Eiffel, France
George Papagiannakis	University of Crete, Greece
Kanta Tachibana	Kogakuin University, Japan
Lars Tingelstad	Norwegian University of Science and Technology, Norway
Petr Vasik	Brno University of Technology, Czech Republic
Yu Zhaoyuan	Nanjing Normal University, China

Additional Reviewers

Arturas Acus
Łukasz Błaszczyk
Roman Byrtus
Emanuele Dalla Torre
Anna Derevianko
Leo Dorst
Ivan Eryganov
Neil Gordon
Jaroslav Hrdina
Jin Huang
Manos Kamarianakis
Ping Li
Pan Lian
Pavel Loučka
Stephen Mann
Stephen Sangwine
Dmitry Shirokov
Jianlin Zhu

Contents

Foundations of Geometric Algebra

The Supergeometric Algebra: The Square Root of the Geometric Algebra

Andrew J. S. Hamilton(✉)

JILA and Department of Astrophysical and Planetary Sciences,
Box 440, U. Colorado Boulder, Boulder, CO 80309, USA
Andrew.Hamilton@colorado.edu
https://jila.colorado.edu/~ajsh/

Abstract. This paper gives a pedagogical account of the Supergeometric Algebra (SGA), the square root of the Geometric Algebra (GA). The fact that a spinor can be treated as a bitcode is emphasized.

Keywords: Supergeometric Algebra · Geometric Algebra · spinors

1 Introduction

It is remarkable that the foundations of the Clifford algebra, or Geometic Algebra (GA), were established a century and a half ago by Grassmann [1,2] and Clifford [3], but it took a David Hestenes [4–7] to berate the physicists that the GA is something they really ought to pay attention to.

I think the Supergeometric Algebra (SGA), the extension of the GA to include spinors, deserves similar close attention by a wider audience. The name follows the common practice of physicists to prepend the word "super" to spinorial extensions of theories. From a strictly mathematical perspective, there's nothing new in this paper. Cartan, who introduced spinors to mathematics in 1913 [8], was thoroughly familiar with everything to do with geometric algebras and spinor algebras [9].

The present paper aims to give a pedagogical introduction to the SGA. A more formal exposition can be found in [10]. The most important single concept I hope to convey is that

$$\boxed{\text{A spinor is a bitcode.}} \tag{1}$$

The second important concept is that, as proved by Brauer & Weyl (1935) [11],

$$\boxed{\text{The Geometric Algebra is the Supergeometric Algebra squared.}} \tag{2}$$

If you are a computer scientist, you should be intrigued by the notion that a spinor is a bitcode. If you are interested in the GA, you should be aware of the fact that there is a natural way to represent objects in the GA with a bitcode.

E. Hitzer et al. (Eds.): ENGAGE 2022, LNCS 13862, pp. 3–15, 2023.
https://doi.org/10.1007/978-3-031-30923-6_1

A prominent application of the SGA is to the fermions and forces of physics. A companion [12] to the present paper shows how the Dirac algebra of spacetime symmetries (the GA of Spin(3, 1)) and the geometric algebra of the group Spin(10), well known as a possible grand unified group, combine as commuting subalgebras of the Spin(11, 1) geometric algebra in 11+1 spacetime dimensions, unifying the four forces of Nature. The paper [12] is based on [13].

For simplicity, the treatment in the present paper takes all dimensions to be spatial. All results generalize to arbitrary dimensions of space and time. For the most part, time dimensions can be treated mathematically as if they were imaginary (with respect to the imaginary i) spatial dimensions.

2 Spinors as a Bitcode

Background. When a gymnast or ballet dancer rotates by one full turn, they return to where they started. Human experience might suggest that this is a law of Nature, that anything rotated by one full turn would necessarily return to its original state. Cartan first showed in 1913 [8] that mathematically there are more fundamental objects, which he called spinors (French spineurs), that require two full turns to return them to their original state. Cartan showed moreover that, within the context of rotations, there is nothing more fundamental than spinors. These properties stem from the topological properties of the rotation group: the usual rotation group (the special orthogonal group SO(N), in N spatial dimensions) is not simply-connected, but it has a double cover (Spin(N)) that is simply-connected. Figure 1 illustrates Dirac's belt trick [14], a well-known demonstration of the non-trivial topological properties of the rotation group.

Dirac in 1928 [15] rediscovered spinors from a physics perspective when he discovered his eponymous equation, a relativistic version of Schrödinger's [16] non-relativistic equation of quantum mechanics. Dirac's equation provided an experimentally successful description of the behavior of the electron, and predicted that an electron should have an antiparticle, a positron, which was discovered in 1932 [17].

With the establishment of the standard model of physics in the 1960s s and 1970s s (see [12] for more), it has become apparent that all known matter (fermions and quarks) is made of spinors, and that all known forces (the three forces of the standard model, plus gravity) emerge from symmetries of spinors. If indeed spinors are so fundamentally plumbed into the laws of Nature, then we humans would do well to pay attention.

Spinors, Vectors, and Rotors. In physics, a spinor is an object of spin $\frac{1}{2}$, whereas a vector is an object of spin 1. Whereas a vector rotates to itself under a rotation by 360°, a spinor requires two turns, 720°, to rotate it back to itself. Spinors and vectors exist in arbitrary dimensions of space, and more generally of spacetime.

If you are familiar with the GA, you may perhaps have heard the idea that a spinor is a rotor. That is not the right way to think about a spinor. A rotor is an element of the group Spin(N) of rotations in N space(time) dimensions. As a

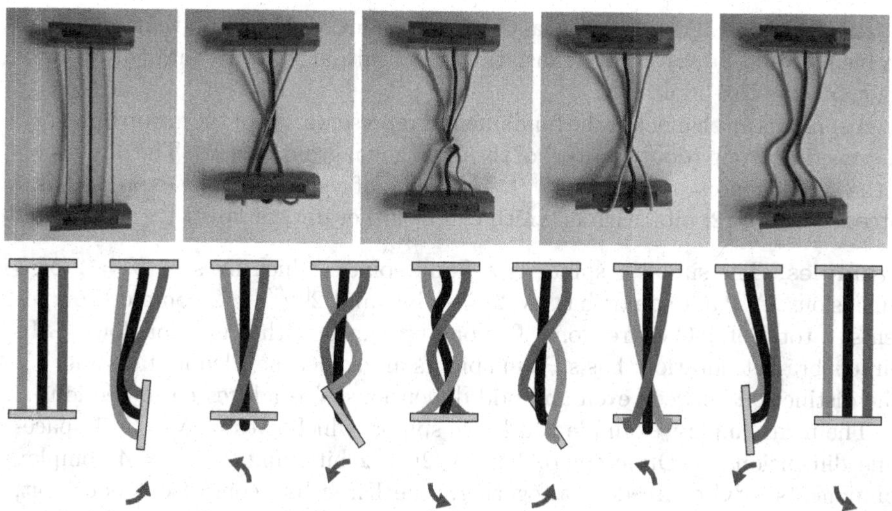

Fig. 1. A version of Dirac's belt trick [14], which illustrates the non-trivial topological properties of the rotation group. The trick demonstrates how an object tethered by ropes to another object gets tangled up when rotated by one full turn, but can be returned to its original state by rotating it a second full turn. The upper row of images are photographs of blocks and hawsers crafted by Tomas Herrera. The graphic in the lower row is by Liberty S. Hamilton.

multivector, a rotor is an element of the even geometric algebra; more specifically, a rotor is an element of the group Spin(N) obtained by exponentiating the bivectors of the GA. Any multivector $\boldsymbol{a}$ in the GA transforms under a rotor R as

$$R: \ \boldsymbol{a} \rightarrow R\boldsymbol{a}\overline{R}\,, \tag{3}$$

where $\overline{R}$ is the reverse (inverse) of R, and since a rotor is a multivector, that is how a rotor transforms. By contrast, a spinor ψ transforms under a rotor R as

$$R: \ \psi \rightarrow R\psi\,. \tag{4}$$

It is true that the transformation law (4) makes it legitimate to conceptualize that a spinor encodes a rotation (a Dirac spinor in 3+1 spacetime dimensions is indeed a "Lorentz gyroscope"), but a spinor is mathematically different from a rotor.

In mathematics, a dimension-n representation of a group is a set of $n \times n$ matrices multiplication of which reproduces the action of the group, along with a set of n-dimensional column vectors upon which the matrices act, rotating the vectors among each other.

A Cartesian vector is an element of the fundamental representation of the group SO(N) of orthogonal rotations in N dimensions. The dimension of the vector representation is N. The index i of a vector x_i runs over $i = 1, ..., N$. I was thrilled to learn this secret in high school, that a vector could be represented as an algebraic object x_i with a Cartesian index i. It meant that geometry problems could be solved by translating them into algebra. I could throw away my geometry textbook.

A spinor is an element of the fundamental representation of the group Spin(N), the covering group (double cover) of the orthogonal group SO(N). The dimension of the spinor representation is $2^{[N/2]}$. The index of a spinor can be expressed as a bitcode with $[N/2]$ bits, each of which can be either up $\uparrow$ or down $\downarrow$.

Examples. The simplest spinor is a Pauli spinor, which lives in $N = 2$ or 3 dimensions. A Pauli spinor has $[N/2] = 1$ bit, and $2^{[N/2]} = 2$ complex components, a total of 4°C of freedom. The one bit can be either up $\uparrow$ or down $\downarrow$. In Dirac's bra-ket notation, basis Pauli spinors are sometimes denoted $|\uparrow\rangle$ and $|\downarrow\rangle$. The distinction between even and odd dimensions N is addressed in Section 5.

The next simplest example is a Dirac spinor, which lives in $N = 3+1$ spacetime dimensions. A Dirac spinor has $[N/2] = 2$ bits, and $2^{[N/2]} = 4$ complex components, 8°C of freedom altogether. The Dirac bits comprise a boost bit, which can be either up $\Uparrow$ or down $\Downarrow$, and a spin bit, which can likewise be either up $\uparrow$ or down $\downarrow$. The spinor is said to be right-handed if the boost and spin bits align, $\Uparrow\uparrow$ or $\Downarrow\downarrow$, left-handed if they anti-align, $\Uparrow\downarrow$ or $\Downarrow\uparrow$. The chiral components of a Dirac spinor, right- or left-handed, are called its Weyl components. Only massless spinors can be purely chiral: a massive spinor, such as an electron, is necessarily a (complex) linear combination of right- and left-handed spinors. Chirality plays a central role in the standard model of physics, in that only the left-handed chiral components of Dirac spinors couple to the weak force: the right-handed components do not feel the weak force.

It has been known since the mid 1970s s [18,19] that each generation of fermions of the standard model organizes elegantly as spinors of the group Spin(10) in $N = 10$ dimensions. The companion paper [12] shows how the standard model and the Dirac algebra can be combined as commuting subalgebras of the Spin(11, 1) geometric algebra in $N = 11+1$ spacetime dimensions. Spinors of Spin(11, 1) have $[N/2] = 6$ bits, and $2^{[N/2]} = 64$ components.

Formalities. How should spinors be thought of geometrically? Start with N-dimensional Euclidean space $\mathbb{R}^N$. Partition the orthonormal basis vectors of Euclidean space into $[N/2]$ pairs, and call them γ_i^+ and γ_i^-, $i = 1, ..., [N/2]$. If the number N of dimensions is odd, one vector, γ_N, remains unpaired (see §5 for more on the GA in odd dimensions). Unlike the GA, the SGA requires a complex structure from the outset, involving a commuting imaginary i which can be identified naturally as the quantum mechanical imaginary (do not confuse the index i with the imaginary i). The grouping of vectors into pairs γ_i^+ and γ_i^- stems from this inevitable intrinsic complex structure. This is a good thing, because quantum mechanics requires a complex structure, which must come from somewhere. Chiral combinations γ_i and $\gamma_{\bar{\imath}}$ (with a barred index $\bar{\imath}$) of the orthonormal basis vectors are defined by

$$\gamma_i \equiv \frac{\gamma_i^+ + i\gamma_i^-}{\sqrt{2}} , \quad \gamma_{\bar{\imath}} \equiv \frac{\gamma_i^+ - i\gamma_i^-}{\sqrt{2}} , \tag{5}$$

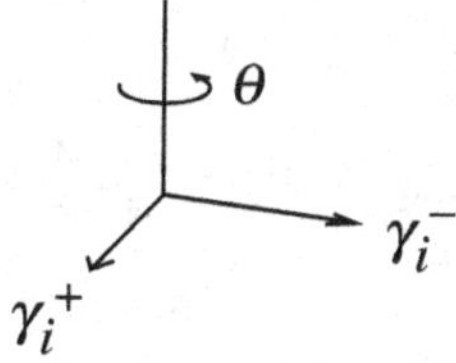

Fig. 2. Right-handed rotation by angle θ in the $\gamma_i^+\gamma_i^-$ plane. The $[N/2]$ conserved charges of Spin(N) are eigenvalues of quantities under rotations in $[N/2]$ planes $\gamma_i^+\gamma_i^-$, $i = 1, ..., [N/2]$. A spinor is a bitcode with $[N/2]$ bits, each of which specifies the corresponding charge of the spinor, either $+\frac{1}{2}$ (↑), or $-\frac{1}{2}$ (↓).

which are normalized so that $\gamma_i \cdot \gamma_{\bar{\imath}} = 1$. The vectors γ_i^+ and $i\gamma_i^-$ can be thought of as, modulo a normalization, the real and imaginary parts of a complex vector γ_i whose complex conjugate is $\gamma_{\bar{\imath}}$.

A pillar of modern physics is Noether's (1918) theorem [20], which states that with each symmetry of a system is associated a conserved charge. Spin(N) has $[N/2]$ conserved charges, which are the eigenvalues of quantities under rotations in each of the $\gamma_i^+\gamma_i^-$ planes, Fig. 2. The $[N/2]$ bits of a spinor specify its $[N/2]$ charges, each of which can be either $+\frac{1}{2}$ (signified up ↑) or $-\frac{1}{2}$ (signified down ↓).

For "ordinary" spatial rotations, the conserved charge is the projection of the angular momentum, or spin, in the $\gamma_i^+\gamma_i^-$ plane (in fundamental units, $\hbar = 1$). However, in other applications of the SGA, the word charge may refer to other conserved charges, such as the conserved charges of the standard model of physics [12].

Under a right-handed rotation by angle θ in the $\gamma_i^+\gamma_i^-$ plane, Fig. 2, the chiral basis vectors γ_i and $\gamma_{\bar{\imath}}$ transform by a phase, Fig. 3,

$$\gamma_i \rightarrow e^{-i\theta}\,\gamma_i\ , \quad \gamma_{\bar{\imath}} \rightarrow e^{i\theta}\,\gamma_{\bar{\imath}}\ . \tag{6}$$

The signs follow the physics convention that a right-handed rotation by angle θ rotates a phase by $e^{-i\theta}$. (If one of the two orthonormal dimensions, say γ_i^-, is a time dimension, then the rotation in the $\gamma_i^+\gamma_i^-$ plane becomes a Lorentz boost, and the transformation (6) becomes

$$\gamma_i \rightarrow e^{\theta}\,\gamma_i\ , \quad \gamma_{\bar{\imath}} \rightarrow e^{-\theta}\,\gamma_{\bar{\imath}}\ .) \tag{7}$$

The transformation (6) identifies the chiral basis vectors γ_i and $\gamma_{\bar{\imath}}$ as having i-charge equal to $+1$ and -1. All other chiral basis vectors, γ_j and $\gamma_{\bar{\jmath}}$ with $j \neq i$, along with the unpaired basis vector γ_N if N is odd, remain unchanged under a rotation in the $\gamma_i^+\gamma_i^-$ plane, so have zero i-charge. The i-charge of a multivector (or tensor of multivectors) can be read off from its covariant chiral indices:

$$i\text{-charge} = \text{number of } i \text{ minus } \bar{\imath} \text{ covariant chiral indices}\ . \tag{8}$$

A spinor ψ,

$$\psi = \psi^a \epsilon_a\ , \tag{9}$$

Fig. 3. The spiral lines track the phase angle $\mp\theta$ of right- and left-handed chiral basis vectors γ_i and $\gamma_{\bar\imath}$ (left two images), Eqs. (6), and $\mp\theta/2$ of basis spinors $\epsilon_{...i...}$ and $\epsilon_{...\bar\imath...}$ with i'th bit up and down (right two images), Eqs. (11), under a rotation by angle θ in the $\gamma_i^+\gamma_i^-$ plane. It takes one full turn, $\theta = 2\pi$, to rotate vectors γ_i and $\gamma_{\bar\imath}$ to themselves, but two full turns, $\theta = 4\pi$, to rotate the spinors $\epsilon_{...i...}$ and $\epsilon_{...\bar\imath...}$ to themselves.

is a complex (with respect to the imaginary i) linear combination of $2^{[N/2]}$ basis spinors ϵ_a. Chiral basis spinors comprise $2^{[N/2]}$ basis spinors ϵ_a,

$$\epsilon_a \equiv \epsilon_{a_1...a_{[N/2]}} \tag{10}$$

where $a = a_1...a_{[N/2]}$ denotes a bitcode of length $[N/2]$. Each bit a_i is either up $\uparrow$ or down $\downarrow$. For example, one of the basis spinors is the all-bit-up basis spinor $\epsilon_{\uparrow\uparrow...\uparrow}$.

Under a right-handed rotation by angle θ in the $\gamma_i^+\gamma_i^-$ plane, basis spinors $\epsilon_{...i...}$ and $\epsilon_{...\bar\imath...}$ with i-bit respectively up and down transform as, Fig. 3,

$$\epsilon_{...i...} \to e^{-i\theta/2}\,\epsilon_{...i...}\ , \quad \epsilon_{...\bar\imath...} \to e^{i\theta/2}\,\epsilon_{...\bar\imath...}\ . \tag{11}$$

The transformation (11) shows that basis spinors $\epsilon_{...i...}$ and $\epsilon_{...\bar\imath...}$ have i-charge respectively $+\frac{1}{2}$ and $-\frac{1}{2}$ in each of its $[N/2]$ bits. The i-charge of a spinor (or tensor of spinors) can be read off from its covariant chiral indices:

$$i\text{-charge} = \tfrac{1}{2}(\text{number of } i \text{ minus } \bar\imath \text{ covariant chiral indices})\ . \tag{12}$$

whereas an orthonormal Cartesian basis vector γ_i^+ or γ_i^- sticks out in one dimension at a time, a basis spinor ϵ_a sticks out in all dimensions at once. This sticking-out-in-all-dimensions-at-once, like a hedgehog, is perhaps one of the things that makes it hard to visualize a spinor. The reason a Cartesian vector can stick out in a single dimension i is that it can be constructed from a tensor product of spinor pairs in which the i'th bit of each of the two spinors points in the same direction, while all bits other than i point in opposite directions, canceling each other out.

3 Spinor Metric

The existence of a metric is fundamental to the GA, and the existence of a spinor metric is similarly fundamental to the SGA. The Euclidean metric δ_{ij} (or Minkowski metric η_{mn}) is that vectorial tensor of rank 2 that remains invariant under SO(N) (or SO(K, M) in $K{+}M$ spacetime dimensions). Similarly, the spinor metric ε_{ba} is that spinor tensor of rank 2 that remains invariant under Spin(N) (or Spin(K, M) in $K+M$ spacetime dimensions).

The scalar product of two spinors χ and ψ can be denoted with a dot,

$$\chi \cdot \psi \,. \tag{13}$$

The fact that the scalar product must be a scalar, therefore carry zero charge, implies that the spinor metric $\varepsilon_{ba} \equiv \boldsymbol{\epsilon}_b \cdot \boldsymbol{\epsilon}_a$ can be non-zero only between basis spinors whose indices are bit-flips of each other, $b = \bar{a}$. Each non-zero component $\varepsilon_{\bar{a}a}$ of the spinor metric equals ± 1, with the sign depending on the component a and the number N of dimensions. See [10] for details.

Whereas the Euclidean (or Minkowski) metric must be symmetric, the spinor metric can be either symmetric or antisymmetric. There prove to be two possible choices for the spinor metric, differing from each other by a factor of the pseudoscalar, denoted ε and ε_{alt}. In Nature, it is Nature that makes the choice. The following chart shows the symmetry of the spinor metric ε or ε_{alt} in N spacetime dimensions:

N(mod 8) :	1	2	3	4	5	6	7	8
ε^2	+	+	−	−	−	−	+	+
$\varepsilon^2_{\text{alt}}$	+	−	−	−	−	+	+	+

(14)

The chart exhibits the well known period-8 Cartan-Bott periodicity [21,22] of geometric algebras.

The chart (14) shows that in 3 or 4 spacetime dimensions, the spinor metric is necessarily antisymmetric. Thus the Pauli metric in $N = 3$ dimensions, and the Dirac metric in $N = 3{+}1$ spacetime dimensions, are necessarily antisymmetric.

4 Column Spinors and Row Spinors

It is not only mathematically correct (in the context of representation theory), but also conceptually helpful, to think of a spinor ψ as a column vector (of dimension $2^{[N/2]}$), and a rotor R as a matrix that acts on the column spinor ψ. More generally, any multivector $\boldsymbol{a}$ can be represented as a $2^{[N/2]} \times 2^{[N/2]}$ matrix that acts by matrix multiplication on a column spinor ψ, yielding $\boldsymbol{a}\psi$.

Associated with any column spinor ψ is a row spinor $\psi\cdot$, equal to the transpose of the column spinor ψ multiplied by the spinor metric tensor ε,

$$\psi \cdot \equiv \psi^{\top} \varepsilon \,. \tag{15}$$

A scalar product $\chi \cdot \psi$ of spinors can be thought of as the matrix product of a row spinor $\chi\cdot$ with a column spinor ψ,

$$\chi \cdot \psi = \chi^{\top} \varepsilon \, \psi \,. \tag{16}$$

The notation $\psi\cdot$ for a row spinor, with a trailing dot symbolizing the spinor metric, is extremely convenient. The dot immediately distinguishes a row spinor from a column spinor; and the dot makes transparent the application of the associative rule to a sequence of products of spinors, Eq. (19). A row spinor $\psi\cdot$ transforms under a rotor R as

$$R: \ \psi\cdot \rightarrow \psi\cdot\overline{R} \,, \tag{17}$$

as follows from the fact that a spinor transforms as (4), and a scalar product of a row and column spinor must be a scalar.

In opposite order, the product of a column spinor ψ and a row spinor $\chi\cdot$ defines their outer product $\psi\chi\cdot$. The outer product transforms under a rotation in the same way (3) as a multivector,

$$R: \ \psi\chi\cdot \equiv \psi\chi^{\top}\varepsilon \rightarrow (R\psi)(R\chi)^{\top}\varepsilon = R(\psi\chi\cdot)\overline{R} \,. \tag{18}$$

Multiplication of outer products satisfies the associative rule

$$(\psi\chi\cdot)(\varphi\xi\cdot) = \psi(\chi\cdot\varphi)\xi\cdot \,, \tag{19}$$

which since $\chi\cdot\varphi$ is a scalar is proportional to the outer product $\psi\xi\cdot$. The associative rule (19) makes it straightforward to simplify long sequences of products of column and row spinors, a process known in quantum field theory as Fierz rearrangement.

A core property of spinors in physics is that they satisfy an exclusion principle. The exclusion principle underlies much of the richness of the behavior of matter at low energy. According to the usual rules of matrix multiplication, a row matrix can multiply a column matrix, yielding a scalar, and a column matrix can multiply a row matrix, yielding a matrix, but a row matrix cannot multiply a row matrix, and a column matrix cannot multiply a column matrix:

$$\begin{pmatrix} \quad & \quad \end{pmatrix}\begin{pmatrix} \\ \\ \end{pmatrix} = (\;) \qquad \text{inner product} = \text{scalar} \,, \tag{20a}$$

$$\begin{pmatrix} \\ \\ \end{pmatrix}\begin{pmatrix} \quad & \quad \end{pmatrix} = \begin{pmatrix} \quad & \quad \\ & \\ & \end{pmatrix} \qquad \text{outer product} = \text{multivector} \,, \tag{20b}$$

$$\begin{pmatrix} \quad & \quad \end{pmatrix}\begin{pmatrix} \quad & \quad \end{pmatrix} = \varnothing \qquad \text{forbidden} \,, \tag{20c}$$

$$\begin{pmatrix} \\ \\ \end{pmatrix}\begin{pmatrix} \\ \\ \end{pmatrix} = \varnothing \qquad \text{forbidden} \,. \tag{20d}$$

These rules resemble the rules for fermionic creation and destruction operators in quantum field theory: creation following destruction is allowed, and destruction following creation is allowed, but creation following creation is forbidden, and destruction following destruction is forbidden. It can be shown that the multiplication rules for row and column spinors indeed reproduce those of fermion creation (row) and destruction (column) operators in quantum field theory.

It would seem that the distinction between column and row spinors, as realized in Nature, is profound.

5 The GA Is the Square of the SGA

Brauer & Weyl (1935) [11] first proved the theorem that the algebra of outer products of spinors is isomorphic to the GA, in any number of even N spacetime dimensions. They used a language familiar to physicists, that of tensors, and representations of groups. [10] gives a proof of the theorem in the language of the GA.

In the notation of the present paper, the Brauer-Weyl isomorphism says that there is an invertible linear mapping between outer products of column and row basis spinors ϵ_a and $\epsilon_b\cdot$ and basis multivectors γ_A of all grades,

$$\epsilon_a \epsilon_b \cdot = c^A_{ab} \gamma_A , \quad \gamma_A = c^{ab}_A \epsilon_a \epsilon_b \cdot , \tag{21}$$

that respects the algebraic structure, that is, it respects addition and multiplication of spinors and multivectors. The outer product is neither symmetric nor antisymmetric in ab. The full set of $2^{N/2} \times 2^{N/2}$ outer products of basis spinors yields the entire 2^N-dimensional geometric algebra.

The simplest example of the Brauer-Weyl isomorphism is the Pauli SGA in $N = 2$ dimensions, where the outer products of the two spinors $\uparrow$ and $\downarrow$ map to the basis multivectors of the GA in 2 dimensions, consisting of one scalar, 2 vectors, and one pseudoscalar, a total of $1+2+1 = 4 = 2^2$ multivectors,

$$\underbrace{1 = (\downarrow\uparrow - \uparrow\downarrow)\cdot}_{\text{1 scalar}} , \quad \underbrace{\gamma_1 = \sqrt{2}\uparrow\uparrow\cdot , \quad \gamma_{\bar{1}} = -\sqrt{2}\downarrow\downarrow\cdot}_{\text{2 vectors}} , \quad \underbrace{I_2 = -i(\downarrow\uparrow + \uparrow\downarrow)\cdot}_{\text{1 pseudoscalar}} . \tag{22}$$

The spinor metric adopted in the algebra (22) is the antisymmetric choice (right column) in the chart (14), which ensures that the algebra is the same as that of the Pauli algebra in $N = 3$ dimensions, Eq. (25).

The natural complex structure of spinors means that spinors live naturally in even spacetime dimensions N. The group Spin(N) on the other hand exists in either even or odd dimensions, and likewise the GA lives in both even and odd dimensions. There are two ways to extend the SGA to odd N dimensions.

The first is to project the odd N-dimensional GA into one lower dimension, by identifying the pseudoscalar I_N of the odd-dimensional GA with the unit multivector (times a phase), whereupon the pseudoscalar I_{N-1} of the one-dimension-lower even-dimensional algebra is promoted to a vector in the N-dimensional algebra.

An example is the Pauli algebra in $N = 3$ dimensions. The orthonormal basis vectors of the Pauli algebra, here denoted γ_1^+, γ_1^-, and γ_3, are commonly denoted σ_i, $i = 1, 2, 3$. The pseudoscalar I_3 of the Pauli algebra, the product of the three vectors, is identified with the imaginary i times the unit scalar 1,

$$I_3 \equiv \gamma_1^+ \gamma_1^- \gamma_3 \,(= \sigma_1 \sigma_2 \sigma_3) = i\,1 . \tag{23}$$

As a result of the identification (23), the pseudoscalar I_2 of the 2-dimensional algebra is promoted to a vector of the 3-dimensional algebra,

$$I_2 \equiv \gamma_1^+ \gamma_1^- \,(= \sigma_1 \sigma_2) = i\gamma_3 \,(= i\sigma_3) . \tag{24}$$

The 3-dimensional geometric algebra differs from the 2-dimensional geometric algebra in that the former possesses a higher level of symmetry: whereas in 2 dimensions there is just one rotation, generated by the bivector $\sigma_1\sigma_2$, in 3 dimensions there are 2 more rotations, generated by the bivectors $\sigma_1\sigma_3$ and $\sigma_2\sigma_3$.

The Pauli SGA in $N = 3$ dimensions (with the standard choice ε of spinor metric, the center column in the chart (14)) is essentially identical to the Pauli SGA (22) in $N = 2$ dimensions, except that the 2D pseudoscalar I_2 is promoted to a vector γ_3, Eq. (24),

$$\underset{\text{1 scalar}}{1 = (\downarrow\uparrow - \uparrow\downarrow)\cdot}\,, \quad \underbrace{\gamma_1 = \sqrt{2}\uparrow\uparrow\cdot\,, \quad \gamma_3 = -(\downarrow\uparrow + \uparrow\downarrow)\cdot\,, \quad \gamma_{\bar{1}} = -\sqrt{2}\downarrow\downarrow\cdot}_{\text{3 vectors}}\,. \tag{25}$$

The rest of the Pauli GA, comprising the 1 pseudoscalar and 3 bivectors, are just i times the 1 scalar and 3 vectors, since the pseudoscalar I_3 has been identified with the imaginary, Eq. (23).

The other way to extend the SGA to odd dimensions is to embed the odd N-dimensional algebra into one higher dimension $N+1$, and to treat the extra vector γ_{N+1} as a scalar. The extra scalar vector γ_{N+1} serves the role of a parity operator (or a time-reversal operator, if one of the dimensions is a time dimension), by virtue of anticommuting with all the original N orthonormal vectors.

6 Conjugation

Spinors have an intrinsic complex structure, and there is a discrete operation, complex conjugation, that converts spinors (and multivectors) into their complex conjugates. The basis spinors $\boldsymbol{\epsilon}_a$ are treated as real, so the complex conjugate of a spinor $\psi \equiv \psi^a \boldsymbol{\epsilon}_a$ is the spinor with complex-conjugated coefficients,

$$\psi^* \equiv (\psi^a)^* \boldsymbol{\epsilon}_a\,. \tag{26}$$

In quantum field theory, complex conjugation turns a spinor into an anti-spinor.

The operation (26) of complex conjugation is not however Lorentz-covariant; under a rotor R, the complex conjugate spinor ψ^* transforms as

$$R:\ \psi^* \to (R\psi)^* = R^*\psi^*\,. \tag{27}$$

The conjugation operator C is introduced to restore Lorentz covariance. The conjugate spinor $\bar{\psi}$ is defined to be the product of the conjugation operator C and the complex conjugate spinor ψ^*,

$$\bar{\psi} \equiv C\psi^*\,. \tag{28}$$

(Do not confuse conjugation with reversion in the GA; the conjugation overbar $\bar{}$ is shorter and thinner than the reversion overbar $\overline{}$.) The conjugation operator C is defined as a Lorentz-invariant operator with the property that commuting it through any rotor R converts the rotor to its complex conjugate, $CR^* = RC$. With the conjugation operator so defined, the conjugate spinor $\bar{\psi}$ transforms under a rotor in the same way as any other spinor,

$$R: \ \bar{\psi} \to C(R\psi)^* = RC\psi^* = R\bar{\psi} \ . \tag{29}$$

The conjugate spinor $\bar{\psi}$ is the antiparticle of the spinor ψ, expressed in a Lorentz-covariant fashion.

In physics, the operation of conjugation is often conflated with the operation of converting a column spinor to a row spinor, so that the conjugate of a spinor ψ is taken to be the conjugate row (or row conjugate) spinor $\bar{\psi}\cdot$. The reason for the conflation is that in quantum field theory a field is a linear combination of creation and destruction operators, and the partner of a fermion destruction operator (column spinor) is the anti-fermion creation operator (conjugate row spinor). However, the two operations are distinct, and it is wise to keep them so. All four operations — fermion creation and destruction, and anti-fermion creation and destruction — occur in quantum field theory.

If all spatial N dimensions are spatial (no time dimensions), then the conjugation operator C coincides with the spinor metric tensor ε. If there is a time dimension, then in the chiral representation the conjugation operator is, up to a phase, the product of the spinor metric and the time vector (or a product of all the time vectors, if there is more than one time dimension). Some texts refer to the spinor metric tensor as the conjugation operator, which I find egregiously confusing.

7 Supersymmetry

The Supergeometric Algebra is *not* the same as the algebra of supersymmetry. The supersymmetry algebra is the extension of the Poincaré algebra to include symmetries generated by spinors. The Poincaré algebra is the algebra of global translations and Lorentz transformations of flat (Minkowski) space. The Poincaré algebra is not the same as the Dirac algebra (the GA in 3+1 spacetime dimensions). The Poincaré and Dirac algebras share the property of having both vector and bivector generators, but the vectors of the Poincaré algebra, the momentum generators P_m, commute, whereas the vectors γ_m of the Dirac algebra anticommute.

In the SGA in 4 spacetime dimensions, the 4 symmetrized outer products of the 2 right-handed with the 2 left-handed spinors yield the 4 chiral basis vectors. The coefficients of the mapping coincide with those of the supersymmetry algebra, which is unsurprising since the mapping must respect the properties of spinors and vectors under Lorentz transformations.

My own idiosyncratic view is that supersymmetry may not be Nature's way. The algebra of outer products of spinors yields the entire geometric algebra, including multivectors of all grades, not just vectors. String theory is apparently a theory not merely of strings (whose worldtubes are generated by bivectors), but of branes of all dimensions (whose worldtubes are generated by multivectors of all grades). Hopefully Nature's way will in due course become apparent to enterprising observers and experimentalists, as has happened in the past.

References

1. Grassmann, H., Ausdehnungslehre, D.: Vollständig und in strenger Form begründet. Enslin, Berlin (1862)
2. Grassmann, H.: Der Ort der Hamilton's chen Quaternionen in der Audehnungslehre. Math. Ann. **12**, 375–386 (1877). https://doi.org/10.1007/BF01444648
3. Clifford, W.K.: Applications of Grassmann's extensive algebra. Am. J. Math. **1**, 350–358 (1878). https://doi.org/10.2307/2369379
4. Hestenes, D.: Space-Time Algebra, Gordon & Breach (1966). https://doi.org/10.1007/978-3-319-18413-5
5. Hestenes, D., Sobczyk, G.: Clifford Algebra to Geometric Calculus. D. Reidel Publishing Company, Dordrecht (1987). https://doi.org/10.1007/978-94-009-6292-7
6. Gull, S., Lasenby, A., Doran, C.: Imaginary numbers are not real–the geometric algebra of spacetime. Found. Phys. **23**, 1175–1201 (1993). https://doi.org/10.1007/BF01883676
7. Lounesto, P.: Clifford Algebras and Spinors, 2nd edn. Cambridge University Press, Cambridge (2001). (London Mathematical Society Lecture Note Series. 286), https://doi.org/10.1017/CBO9780511526022
8. Cartan, E.: Les groupes projectifs qui ne laissent invariante aucune multiplicité plane. Bulletin Société Mathématique de France **41**, 53–96 (1913). https://doi.org/10.24033/bsmf.916
9. Cartan, É.: Leçons sur la théorie des spineurs (Hermann & Cie, Paris, 1938); English translation The Theory of Spinors. MIT Press, Cambridge, MA (1966)
10. Hamilton, A.J.S.: The supergeometric algebra. Advances in Applied Clifford Algebras, submitted (2022)
11. Brauer, R., Weyl, H.: Spinors in n dimensions. Am. J. Math. **57**, 425–449 (1935). https://doi.org/10.2307/2371218
12. Hamilton, A.J.S.: The supergeometric algebra as the language of physics. In: International Conference of Advanced Computational Applications of Geometric Algebra (ICACAGA), Denver, Colorado, October 2022, submitted
13. Hamilton, A.J.S., McMaken, T.C.: Unification of the four forces in the Spin(11,1) geometric algebra. J. Phys. A (2022). submitted
14. Staley, M.: Understanding quaternions and the Dirac belt trick. Eur. J. Phys. **31**, 467–478 (2010). https://doi.org/10.1088/0143-0807/31/3/004
15. Dirac, P.A.M.: The quantum theory of the electron. Proc. Roy. Soc. **117**, 610–624 (1928). https://doi.org/10.1098/rspa.1928.0023
16. Schrödinger, E.: An undulatory theory of the mechanics of atoms and molecules. Phys. Rev. **28**, 1049–1070 (1926). https://doi.org/10.1103/PhysRev.28.1049
17. Anderson, C.D.: The positive electron. Phys. Rev. **43**, 491–494 (1933). https://doi.org/10.1103/PhysRev.43.491
18. Fritzsch, H., Minkowski, P.: Unified interactions of leptons and hadrons. Ann. Phys. **93**, 193–266 (1975). https://doi.org/10.1016/0003-4916(75)90211-0
19. Wilczek, F.: SO(10) marshals the particles. Nature **394**, 15 (1998). https://doi.org/10.1038/27761

20. Noether, E.: Invariante Variationsprobleme. Nachr. D. König. Gesellsch. D. Wiss. Zu Göttingen, Math-phys. Klasse **1918**, 235–257 (1918)
21. Study, E., Cartan, É.: Nombres complexes. Encyclopédie des sciences mathématiques; tome 1, volume 1, fascicule 3, pp. 353–411 (1908)
22. Bott, R.: The periodicity theorem for the classical groups and some of its applications. Adv. Math. **4**, 353–411 (1970). https://doi.org/10.1016/0001-8708(70)90030-7

Calculation of the Exponential in Arbitrary $Cl_{p,q}$ Clifford Algebra

Arturas Acus[1(✉)] and Adolfas Dargys[2]

[1] Institute of Theoretical Physics and Astronomy, Vilnius University, Saulėtekio 3, 10257 Vilnius, Lithuania
arturas.acus@tfai.vu.lt

[2] Semiconductor Physics Institute, Center for Physical Sciences and Technology, Saulėtekio 3, 10257 Vilnius, Lithuania
adolfas.dargys@ftmc.lt

Abstract. Formulas to calculate multivector exponentials in a base-free representation and in a given orthogonal basis are presented for an arbitrary Clifford geometric algebra $Cl_{p,q}$. The formulas are based on the analysis of roots of the characteristic polynomial of a multivector exponent. Elaborate examples how to use the formulas in practice are presented. The results may be useful in theory of quantum circuits or in the problems of analysis of evolution of the entangled quantum states.

Keywords: Clifford (geometric) algebra · exponentials of Clifford numbers · computer-aided theory

1 Introduction and Notation

Mathematical models of physical, economical, biological, etc. processes often require computation of exponential of matrix. Since in many applications the matrices can be replaced by multivectors (MV), the exponential of MV [13,14,19,20] in geometric (Clifford) algebras has a wide range of applications as well.

Exponential of matrix can be computed by a number of different ways [6,11,12,16,28]. The review article [21] presents twenty methods[1] related to the approximate (finite precision) methods only. According to [21], our approach in this paper can be identified as METHOD 8 and falls into the class of *polynomial methods*, except that here we provide explicit and exact formulas for the basis expansion coefficients instead of recursive approximation. The polynomial methods [21] are known to have $O(n^4)$ complexity and, therefore, are prohibitively expensive except for small n. As far as the exact (closed form) formulas for exponentials and other functions are concerned, most of works deal either with

[1] The article is named "Nineteen dubious ways to compute the exponential of a matrix, twenty-five years later". One more method was added in the revised version of the original article (published in 1978), however, authors wanted to preserve the article title. The next article update is planned in 2028.

E. Hitzer et al. (Eds.): ENGAGE 2022, LNCS 13862, pp. 16–27, 2023.
https://doi.org/10.1007/978-3-031-30923-6_2

the low dimensional cases [11,12,28] (dimensions 5 and 6 are already causing problems [16]) or with matrices that are representations of some Lie groups [22] or, alternatively, have some other special symmetries [6].

In the context of geometric algebra (GA) there often appears a need to compute the rotor, which is an exponential of bivector. The simplest half-angle rotors are related to trigonometric and hyperbolic functions. The GA exponential of an arbitrary bivector can be computed using the method of invariant decomposition [23], where the bivector is decomposed into commuting orthogonal 2-blades, exponentiation of which are more or less straightforward. For low dimensional cases other decomposition techniques can be applied as well [10,17].

When dealing with exponentials of pure bivector $\mathcal{A}$ one should always keep in mind that in general they do not form a group. For example, there are elements of $Spin_+(2,2)$ which can't be written in the form $\pm e^{\mathcal{A}}$. Also $SO_+(1,3)$ contains elements, which are not exponentials of bivectors [19], p224.

For $n \leq 3$ explicit formulas for computation of general exponentials [3,8,9] and all of square roots [2,7] are known. The formulas for low dimensional algebras are faster and more easy to implement.

In this paper the explicit formula to calculate the exponential function of a general MV in an arbitrary $Cl_{p,q}$ is presented. In Sect. 2 the methods to generate characteristic polynomials in $Cl_{p,q}$ algebras characterized by arbitrary signature $\{p,q\}$ and vector space dimension $n = p+q$ are discussed. The method of calculation of the exponential is presented in Sect. 3. In Sect. 4 we demonstrate that the obtained GA exponentials may be used to find the elementary and special GA functions. Below, the notation used in the paper is described briefly. For those readers who are unfamiliar with Clifford geometric algebras we recommend an excellent textbook by Lounesto [19].

In the orthonormalized basis used here the geometric product of basis vectors $\mathbf{e}_i$ and $\mathbf{e}_j$ satisfy [19] the anti-commutation relation, $\mathbf{e}_i\mathbf{e}_j + \mathbf{e}_j\mathbf{e}_i = \pm 2\delta_{ij}$. The number of subscripts indicates the grade. For a mixed signature $Cl_{p,q}$ algebra the squares of basis vectors, correspondingly, are $\mathbf{e}_i^2 = +1$ and $\mathbf{e}_j^2 = -1$, where $i = 1,2,\ldots,p$ and $j = p+1, p+2, \ldots, p+q$. The sum $n = p+q$ is the dimension of the vector space. The general MV is expressed as

$$\mathsf{A} = a_0 + \sum_i a_i \mathbf{e}_i + \sum_{i<j} a_{ij}\mathbf{e}_{ij} + \cdots + a_{1\cdots n}\mathbf{e}_{1\cdots n} = a_0 + \sum_J^{2^n-1} a_J \mathbf{e}_J, \tag{1}$$

where a_i, $a_{ij\ldots}$ are the real coefficients. The ordered set of indices will be denoted by a single capital letter J referred to as a multi-index. Note, that in the multi-index representation the scalar is deliberately excluded from summation as indicated by the upper range $2^n - 1$ in the sum in the last expression.

We shall need three grade involutions: the reversion (e.g., $\widetilde{\mathbf{e}_{12}} = \mathbf{e}_{21} = -\mathbf{e}_{12}$), the grade inverse (e.g., $\widehat{\mathbf{e}_{123}} = -\mathbf{e}_{123}$) and the Clifford conjugation ($\widetilde{\widehat{\mathbf{e}_{123}}} = \mathbf{e}_{123}$). Also we shall need the Hermitian conjugate MV $\mathsf{A}^\dagger$ and non-zero grade negation (see Table 1) operation denoted by overline $\overline{\mathsf{A}}$. The MV Hermitian conjugation expressed for basis elements $\mathbf{e}_J$ in both real and complex GAs can be written as [20,25]

$$\mathsf{A}^{\dagger} = a_0^* + a_1^* \mathbf{e}_1^{-1} + \cdots + a_{12}^* \mathbf{e}_{12}^{-1} + \cdots + a_{123}^* \mathbf{e}_{123}^{-1} \cdots = a_0^* + \sum_J a_J^* \mathbf{e}_J^{-1}, \tag{2}$$

where a_J^* s the complex conjugated J-th coefficient and $\mathbf{e}_J^{-1}$ denotes inverse basis element, $\mathbf{e}_J^{-1}\mathbf{e}_J = 1$.

2 MV Characteristic Polynomial and Equation

The algorithm to calculate the exponential and associated functions presented below is based on a characteristic polynomial. There is a number of methods adapted to MVs, for example, based on MV determinant, recursive Faddeev-LeVerrier method adapted to GA and the method related to Bell polynomials [1, 15, 26]. In this section these methods are briefly summarized.

Every MV $\mathsf{A} \in Cl_{p,q}$ has a characteristic polynomial $\chi_{\mathsf{A}}(\lambda)$ of degree d in $\mathbb{R}$, where $d = 2^{\lceil \frac{n}{2} \rceil}$ is the integer, $n = p + q$. In particular, $d = 2^{n/2}$ if n is even and $d = 2^{(n+1)/2}$ if n is odd. The integer d may be also interpreted as a dimension of real or complex matrix representation of Clifford algebra in the 8-fold periodicity table [19]. The characteristic polynomial [1, 15, 26, 27] is defined by

$$\chi_{\mathsf{A}}(\lambda) = -\mathrm{Det}(\lambda - \mathsf{A}) = \sum_{k=0}^{d} C_{(d-k)}(\mathsf{A})\, \lambda^k. \tag{3}$$

The variable in the characteristic polynomial will be denoted by λ and the roots of $\chi_{\mathsf{A}}(\lambda) = 0$ (called the characteristic equation) by λ_i, respectively. For real GA the coefficients $C_{(k)} \equiv C_{(k)}(\mathsf{A})$ are real. They depend on a selected GA and MV A. The coefficient at the highest power of λ is always assumed $C_{(0)} = -1$. The coefficient $C_{(1)}(\mathsf{A})$ represents MV trace, $C_{(1)}(\mathsf{A}) = \mathrm{Tr}(\mathsf{A}) = d\,\langle \mathsf{A} \rangle_0$, where $\langle \mathsf{A} \rangle_0$ is the scalar part of MV in (1), i.e. $\langle \mathsf{A} \rangle_0 = a_0$. The coefficient $C_{(d)}(\mathsf{A})$ is related to MV determinant $\mathrm{Det}\mathsf{A} = -C_{(d)}(\mathsf{A})$.

Table 1. Optimized expressions for determinant of MV A in low dimensional GAs, $n \le 6$. The overbar denotes a negation of all grades except of the scalar, $\overline{\mathsf{A}} := 2\langle \mathsf{A} \rangle_0 - \mathsf{A}$.

$Cl_{p,q}$	$\mathrm{Det}(\mathsf{A})$
$p + q = 1, 2$	$\mathsf{A}\bar{\mathsf{A}}$
$p + q = 3, 4$	$\frac{1}{3}(\mathsf{A}\mathsf{A}\overline{\mathsf{A}\mathsf{A}} + 2\mathsf{A}\overline{\bar{\mathsf{A}}\,\overline{\bar{\mathsf{A}}\bar{\mathsf{A}}}})$
$p + q = 5, 6$	$\frac{1}{3}(\mathsf{H}\mathsf{H}\overline{\mathsf{H}\mathsf{H}} + 2\mathsf{H}\overline{\bar{\mathsf{H}}\,\overline{\bar{\mathsf{H}}\bar{\mathsf{H}}}})$ with $\mathsf{H} = \mathsf{A}\widetilde{\mathsf{A}}$

Table 1 shows how the MV determinant can be calculated in the low dimensional ($n \le 6$) GAs. This table may be used to find the coefficients $C_{(k)}(\mathsf{A})$ in the characteristic polynomial (3). For a concrete algebra it is enough to replace the products of A's in the Table 1 by products of $(\lambda - \mathsf{A})$.

In Faddeev-Leverrier method [18,27] the coefficients $C_{(k)}(\mathsf{A})$ in polynomial (3) are calculated recursively, beginning from $C_{(1)}(\mathsf{A})$ and ending with $C_{(d)}(\mathsf{A})$. We start from a multivector $\mathsf{A}_{(1)}$ by setting $\mathsf{A}_{(1)} = \mathsf{A}$. Then we compute the coefficient $C_{(k)}(\mathsf{A}) = \frac{d}{k}\langle \mathsf{A}_{(k)}\rangle_0$ and in the next step the new MV $\mathsf{A}_{(k+1)} = \mathsf{A}(\mathsf{A}_{(k)} - C_{(k)})$, where the product on the rhs is a geometric product:

$$\begin{aligned} \mathsf{A}_{(1)} = \mathsf{A} &\to C_{(1)}(\mathsf{A}) = \tfrac{d}{1}\langle \mathsf{A}_{(1)}\rangle_0, \\ \mathsf{A}_{(2)} = \mathsf{A}\big(\mathsf{A}_{(1)} - C_{(1)}\big) &\to C_{(2)}(\mathsf{A}) = \tfrac{d}{2}\langle \mathsf{A}_{(2)}\rangle_0, \\ &\vdots \\ \mathsf{A}_{(d)} = \mathsf{A}\big(\mathsf{A}_{(d-1)} - C_{(d-1)}\big) &\to C_{(d)}(\mathsf{A}) = \tfrac{d}{d}\langle \mathsf{A}_{(d-1)}\rangle_0. \end{aligned} \tag{4}$$

The determinant of A then is $\mathrm{Det}(\mathsf{A}) = -\mathsf{A}_{(d)} = -C_{(d)} = \mathsf{A}\big(\mathsf{A}_{(d-1)} - C_{(d-1)}\big)$.

The coefficients of characteristic equation satisfy the following properties

$$\frac{\partial C_{(k)}(t\mathsf{A})}{\partial t} = kt^{k-1}C_{(k)}(t\mathsf{A}), \qquad \frac{\partial C_{(1)}(t\mathsf{A}^k)}{\partial t} = kt^{k-1}C_{(1)}(t\mathsf{A}^k), \tag{5}$$

where t is a scalar parameter. We shall need (5) in the proofs of theorems.

In the matrix theory a minimal polynomial $\mu_A(\lambda)$ establishes the conditions of diagonalizability of matrix A that represents the MV A. In particular for $n > 2$, it is well-known that matrix is diagonalizable if and only if the minimal polynomial of the matrix does not have multiple roots, i.e. when the minimal polynomial is a product of distinct linear factors. It is also well-known that the minimal polynomial divides the characteristic polynomial. This implies that if roots of the characteristic equation are all different, then matrix/MV is diagonalizable. The polynomial $\mu_{\mathsf{A}}(\lambda)$ can be defined for MV as well. An algorithm on how to compute the minimal polynomial without any reference to matrix representation of the MV is given in Appendix A.

3 MV Exponentials in $Cl_{p,q}$ Algebra

3.1 Exponential of MV in Coordinate (Orthogonal Basis) Form

Theorem 1 (Exponential in coordinate form). *The exponential of a general MV A given by Eq. (1) in $Cl_{p,q}$ is the multivector*

$$\exp(\mathsf{A}) = \frac{1}{d}\sum_{i=1}^{d} \mathrm{e}^{\lambda_i}\left(1 + \sum_{J}^{2^n-1} \mathbf{e}_J \frac{\sum_{m=0}^{d-2}\lambda_i^m \sum_{k=0}^{d-m-2} C_{(k)}(\mathsf{A})\, C_{(1)}(\mathbf{e}_J^{\dagger}\mathsf{A}^{d-k-m-1})}{\sum_{r=0}^{d-1}(r+1)\, C_{(d-r-1)}(\mathsf{A})\,\lambda_i^r}\right) \tag{6}$$

$$= \frac{1}{d}\sum_{i=1}^{d} \exp(\lambda_i)\Big(1 + \sum_{J}^{2^n-1} \mathbf{e}_J\, b_J(\lambda_i)\Big), \qquad b_J(\lambda_i) \in \mathbb{R}, \mathbb{C}. \tag{7}$$

Here λ_i and λ_i^j denotes, respectively, the root of a characteristic equation and the root raised to power j. The (first) sum in (6) is over all roots λ_i of characteristic equation $\chi_{\mathsf{A}}(\lambda) = 0$, where $\chi_{\mathsf{A}}(\lambda)$ is the characteristic polynomial of MV

A *expressed as* $\chi_{\mathsf{A}}(\lambda) = \sum_{i=0}^{d} C_{(d-i)}(\mathsf{A})\,\lambda^i$. *The symbol* $C_{(1)}(\mathbf{e}_J^\dagger \mathsf{A}^k) = d\,\langle \mathbf{e}_J^\dagger \mathsf{A}^k\rangle_0$ *denotes the first coefficient (the coefficient at* λ^{d-1}*) in the characteristic polynomial that consists of geometric product of the Hermitian conjugate basis element* $\mathbf{e}_J^\dagger$ *and k-th power of initial MV:* $\mathbf{e}_J^\dagger \mathsf{A}^k = \mathbf{e}_J^\dagger \underbrace{\mathsf{A}\mathsf{A}\cdots\mathsf{A}}_{k\ terms}$.

Note, because the roots of characteristic equation in general are the complex numbers, the individual terms in sums are complex. However, the result $\exp(\mathsf{A})$ *always simplifies to a real Clifford number if GA is real.*

Proof. Using computer algebra package [4] we first checked the expression was valid for dimensions $n \le 6$. For general n the formula can be proved by using formula (9) from Theorem 2 and noting that projection coefficient onto basis element $\mathbf{e}_J$ can simply be written as $\mathrm{Tr}(\mathbf{e}_J^\dagger \mathsf{A}^r) = C_{(1)}(\mathbf{e}_J^\dagger \mathsf{A}^r)$. □

Example 1. *Exponential of generic MV in* $Cl_{0,3}$ *(different roots).* Let's compute the exponential of $\mathsf{A} = 8 - 6\mathbf{e}_2 - 9\mathbf{e}_3 + 5\mathbf{e}_{12} - 5\mathbf{e}_{13} + 6\mathbf{e}_{23} - 4\mathbf{e}_{123}$ with Eq. (6). We find $d = 4$. Computation of coefficients of the characteristic polynomial $\chi_{\mathsf{A}}(\lambda) = C_{(4)}(\mathsf{A}) + C_{(3)}(\mathsf{A})\lambda + C_{(2)}(\mathsf{A})\lambda^2 + C_{(1)}(\mathsf{A})\lambda^3 + C_{(0)}(\mathsf{A})\lambda^4$ yields $C_{(0)}(\mathsf{A}) = -1$, $C_{(1)}(\mathsf{A}) = 32$, $C_{(2)}(\mathsf{A}) = -758$, $C_{(3)}(\mathsf{A}) = 10432$, $C_{(4)}(\mathsf{A}) = -72693$. The characteristic equation $\chi_{\mathsf{A}}(\lambda) = 0$ then becomes $-72693 + 10432\lambda - 758\lambda^2 + 32\lambda^3 - \lambda^4 = 0$, which has four different roots: $\lambda_1 = 12 - \mathrm{i}\sqrt{53}$, $\lambda_2 = 12 + \mathrm{i}\sqrt{53}$, $\lambda_3 = 4 - \mathrm{i}\sqrt{353}$, $\lambda_4 = 4 + \mathrm{i}\sqrt{353}$. This means that the MV is diagonalizable. For every multi-index J and each root λ_i we have to compute coefficients in Eq.(7),

$$b_J(\lambda_i) = \frac{-\lambda_i^2 C_{(1)}(\mathbf{e}_J^\dagger \mathsf{A}) + \lambda_i\big(32C_{(1)}(\mathbf{e}_J^\dagger \mathsf{A}) - C_{(1)}(\mathbf{e}_J^\dagger \mathsf{A}^2)\big) - 758C_{(1)}(\mathbf{e}_J^\dagger \mathsf{A}) + 32C_{(1)}(\mathbf{e}_J^\dagger \mathsf{A}^2) - C_{(1)}(\mathbf{e}_J^\dagger \mathsf{A}^3)}{-4\lambda_i^3 + 96\lambda_i^2 - 1516\lambda_i + 10432},$$

where we still have to substitute the coefficients $C_{(1)}(\mathbf{e}_J^\dagger \mathsf{A}^k)$

$C_{(1)}(\mathbf{e}_J^\dagger \mathsf{A}^k)$	$\mathbf{e}_{J=1}^\dagger$	$\mathbf{e}_{J=2}^\dagger$	$\mathbf{e}_{J=3}^\dagger$	$\mathbf{e}_{J=12}^\dagger$	$\mathbf{e}_{J=13}^\dagger$	$\mathbf{e}_{J=23}^\dagger$	$\mathbf{e}_{J=123}^\dagger$
$k=1$	0	-24	-36	20	-20	24	-16
$k=2$	192	-224	-416	32	-128	384	-856
$k=3$	8208	5952	5508	-11572	7468	888	-7984

that are different for each multi-index J. The Hermite conjugate elements are $\mathbf{e}_J^\dagger = \{-\mathbf{e}_1, -\mathbf{e}_2, -\mathbf{e}_3, -\mathbf{e}_{12}, -\mathbf{e}_{13}, -\mathbf{e}_{23}, \mathbf{e}_{123}\}$. After substituting all computed quantities into (6) we finally get

$$\begin{aligned}
\exp(\mathsf{A}) &= \frac{1}{2}\mathrm{e}^4\left(\mathrm{e}^8\cos\alpha + \cos\beta\right) + \left(\frac{3}{\alpha}\mathrm{e}^{12}\sin\alpha - \frac{3}{\beta}\mathrm{e}^4\sin\beta\right)\mathbf{e}_1 \\
&+ \left(-\frac{1}{2\alpha}\mathrm{e}^{12}\sin\alpha - \frac{11}{2\beta}\mathrm{e}^4\sin\beta\right)\mathbf{e}_2 + \left(-\frac{2}{\alpha}\mathrm{e}^{12}\sin\alpha - \frac{7}{\beta}\mathrm{e}^4\sin\beta\right)\mathbf{e}_3 \\
&+ \left(-\frac{2}{\alpha}\mathrm{e}^{12}\sin\alpha + \frac{7}{\beta}\mathrm{e}^4\sin\beta\right)\mathbf{e}_{12} + \left(\frac{1}{2\alpha}\mathrm{e}^{12}\sin\alpha - \frac{11}{2\beta}\mathrm{e}^4\sin\beta\right)\mathbf{e}_{13} \\
&+ \left(\frac{3}{\alpha}\mathrm{e}^{12}\sin\alpha + \frac{3}{\beta}\mathrm{e}^4\sin\beta\right)\mathbf{e}_{23} + \frac{1}{2}\mathrm{e}^4\left(\cos\beta - \mathrm{e}^8\cos\alpha\right)\mathbf{e}_{123}, \qquad (8)
\end{aligned}$$

where $\alpha = \sqrt{53}$ and $\beta = \sqrt{353}$.

3.2 Exponential in Basis-Free Form

The basis-free exponential follows from Eq. (6) after addition of terms over the multi-index J.

Theorem 2 (MV exponential in basis-free form). *In $Cl_{p,q}$ algebra the exponential of a general MV A of Eq. (1) can be computed by following formulas*

$$\exp(\mathsf{A}) = \sum_{i=1}^{d} \exp(\lambda_i)\, \beta(\lambda_i) \sum_{m=0}^{d-1} \Big(\sum_{k=0}^{d-m-1} \lambda_i^k C_{(d-k-m-1)}(\mathsf{A}) \Big) \mathsf{A}^m \tag{9}$$

$$= \sum_{i=1}^{d} \exp(\lambda_i) \Big(\frac{1}{d} + \beta(\lambda_i) \sum_{m=0}^{d-2} \Big(\sum_{k=0}^{d-m-2} \lambda_i^k C_{(d-k-m-2)}(\mathsf{A}) \Big) \langle \mathsf{A}^{m+1} \rangle_{-0} \Big) \tag{10}$$

$$= \sum_{i=1}^{d} \exp(\lambda_i) \Big(\frac{1}{d} + \beta(\lambda_i) \mathsf{B}(\lambda_i) \Big), \quad \beta(\lambda_i) = \frac{1}{\sum_{j=0}^{d-1} (j+1)\, C_{(d-j-1)}(\mathsf{A})\, \lambda_i^j}\,. \tag{11}$$

The expression $\langle \mathsf{A}^{m+1} \rangle_{-0} \equiv \frac{1}{2}(\mathsf{A}^{m+1} - \overline{\mathsf{A}^{m+1}})$ indicates that all grades of multivector A^{m+1} are included except of the grade-0, because the scalar part is simply a sum of exponents of eigenvalues divided by d.

The formula (9) has some similarity with exponential of square matrix in [12].

Proof. We will prove basis-free formula (9) by checking the defining equation

$$\frac{\partial \exp(\mathsf{A}t)}{\partial t}\bigg|_{t=1} = \mathsf{A} \exp(\mathsf{A}) = \exp(\mathsf{A})\mathsf{A}, \tag{12}$$

where A is independent of a scalar parameter t. For this purpose we shall verify that the expression (9) for exponential in the Theorem 2 satisfies the identity (12).

First, using properties of characteristic coefficients in (5) and noting that the replacement $\mathsf{A} \to \mathsf{A}t$ implies $\lambda_i \to \lambda_i t$, and performing differentiation $\frac{\partial \exp(\mathsf{A}t)}{\partial t}\Big|_{t=1}$ we obtain that $\exp(\lambda_i)$ in the right hand side of (9) (and also of (6)) after differentiation is replaced by $\lambda_i \exp(\lambda_i)$,

$$\frac{\partial \exp(\mathsf{A}t)}{\partial t}\bigg|_{t=1} = \sum_{i=1}^{d} \lambda_i \exp(\lambda_i) \beta(\lambda_i) \sum_{m=0}^{d-1} \Big(\sum_{k=0}^{d-m-1} \lambda_i^k C_{(d-k-m-1)}(\mathsf{A}) \Big) \mathsf{A}^m, \tag{13}$$

where the weight factor $\beta(\lambda_i)$ further plays no role in the proof. Next, we multiply the basis-free formula (9) by A

$$\mathsf{A} \exp(\mathsf{A}) = \sum_{i=1}^{d} \exp(\lambda_i)\, \beta(\lambda_i) \sum_{m=0}^{d-1} \Big(\sum_{k=0}^{d-m-1} \lambda_i^k C_{(d-k-m-1)}(\mathsf{A}) \Big) \mathsf{A}^{m+1}, \tag{14}$$

and subtract the second equation from the first for each fixed root λ_i, i.e. temporary ignore the summation over roots,

$$\left(\frac{\partial\exp(\mathsf{A}t)}{\partial t}\bigg|_{t=1} - \mathsf{A}\exp(\mathsf{A})\right)\bigg|_{\lambda_i} = \exp(\lambda_i)\,\beta(\lambda_i)\Big(\sum_{k=1}^{d}\lambda_i^k C_{(d-k)}(\mathsf{A}) - \mathsf{A}^k C_{(d-k)}(\mathsf{A})\Big)$$
$$= \exp(\lambda_i)\,\beta(\lambda_i)\Big((\lambda_i^d - \mathsf{A}^d)C_{(0)}(\mathsf{A}) + \cdots + (\lambda_i - \mathsf{A})C_{(d-1)}(\mathsf{A})\Big)\,. \tag{15}$$

Using the Cayley-Hamilton relation for A, which follow from (4),

$$\sum_{k=0}^{d}\mathsf{A}^k C_{(d-k)}(\mathsf{A}) = \mathsf{A}^d C_{(0)}(\mathsf{A}) + \mathsf{A}^{d-1}C_{(1)}(\mathsf{A}) + \cdots + C_{(d)}(\mathsf{A}) = 0,$$

and the same relation for λ_i^d, we solve for the highest powers A^d and λ_i^d, and substitute them into the difference formula (15). As a result, after expansion we obtain zero. □

Example 2. *Exponential of MV in $Cl_{4,0}$ (multiple and zero eigenvalue).* Let's compute the exponential of $\mathsf{A} = -4 - \mathbf{e}_1 - \mathbf{e}_2 - \mathbf{e}_3 - \mathbf{e}_4 - 2\sqrt{3}\mathbf{e}_{1234}$ with basis-free formula (10). Using Table 1 one can easily verify that $\mathrm{Det}(\mathsf{A}) = 0$. For algebra $Cl_{4,0}$ we find $d = 4$. The characteristic polynomial is $\chi_{\mathsf{A}}(\lambda) = C_{(4)}(\mathsf{A}) + C_{(3)}(\mathsf{A})\lambda + C_{(2)}(\mathsf{A})\lambda^2 + C_{(1)}(\mathsf{A})\lambda^3 + C_{(0)}(\mathsf{A})\lambda^4 = -64\lambda^2 - 16\lambda^3 - \lambda^4 = -\lambda^2(8+\lambda)^2$. The roots are $\lambda_1 = 0, \lambda_2 = 0, \lambda_3 = -8, \lambda_4 = -8$. Because multiple roots appear, we have to compute a minimal polynomial of MV A (see Appendix A), which is $\mu_{\mathsf{A}}(\lambda) = \lambda(8+\lambda)$. Since $\mu_{\mathsf{A}}(\lambda)$ has only linear factors, the MV is diagonalizable, and the formula for $\mu_{\mathsf{A}}(\lambda)$ can be applied without modification. It is also easy to verify that the minimal polynomial divides the characteristic polynomial, $\chi_{\mathsf{A}}(\lambda)/\mu_{\mathsf{A}}(\lambda) = \frac{-\lambda^2(8+\lambda)^2}{\lambda(8+\lambda)} = -\lambda(8+\lambda)$. This confirms the property that non-repeating roots of a characteristic polynomial are sufficient but not necessary criterion of MV diagonalizability. Then, we have

$$\beta(\lambda_i)\mathsf{B}(\lambda_i) = \frac{1}{\sum_{j=0}^{d-1}(j+1)\,C_{(d-j-1)}(\mathsf{A})\,\lambda_i^j}\sum_{m=0}^{d-2}\sum_{k=0}^{d-m-2}\lambda_i^k C_{(d-k-m-2)}(\mathsf{A})\,\langle\mathsf{A}^{m+1}\rangle_{-0}$$
$$= \frac{8+\lambda_i}{4\lambda_i(4+\lambda_i)}\langle\mathsf{A}\rangle_{-0} + \frac{16+\lambda_i}{4\lambda_i(4+\lambda_i)(8+\lambda_i)}\langle\mathsf{A}^2\rangle_{-0} + \frac{1}{4\lambda_i(4+\lambda_i)(8+\lambda_i)}\langle\mathsf{A}^3\rangle_{-0}$$
$$= -\frac{1}{\lambda_i+4} - \frac{1}{4(\lambda_i+4)}\mathbf{e}_1 - \frac{1}{4(\lambda_i+4)}\mathbf{e}_2 - \frac{1}{4(\lambda_i+4)}\mathbf{e}_3 - \frac{1}{4(\lambda_i+4)}\mathbf{e}_4 - \frac{\sqrt{3}}{2\lambda_i+8}\mathbf{e}_{1234}. \tag{16}$$

From the middle line one may suppose that the sum over roots would yield division by zero due to zero denominators. The last line, however, demonstrates that this is not the case, since after collecting terms at basis elements we see that all potential zeroes in the denominators have been cancelled. Unfortunately, the cancellation would not occur if the MV were non-diagonalizable. Lastly, after performing summation $\sum_{i=1}^{d}\exp(\lambda_i)\big(\frac{1}{d} + \beta(\lambda_i)\mathsf{B}(\lambda_i)\big)$ over complete set of roots $\{\lambda_1, \lambda_2, \lambda_3, \lambda_4\} = \{0, 0, -8, -8\}$ with exponent weight factor $\exp(\lambda_i)$, which can be replaced by any other function or transformation (see Sect. 4) we obtain

$$\exp(\mathsf{A}) = \frac{1+\mathrm{e}^8}{2\mathrm{e}^8} + \frac{1-\mathrm{e}^8}{8\mathrm{e}^8}\big(\mathbf{e}_1 + \mathbf{e}_2 + \mathbf{e}_3 + \mathbf{e}_4 - 2\sqrt{3}\mathbf{e}_{1234}\big).$$

3.3 Making the Answer Real

Formulas (6) and (9) include summation over (in general complex valued) roots of characteristic polynomial, therefore, formally the result is a complex number. Here we are dealing with real Clifford algebras having real coefficients, consequently, a pure imaginary part, or numbers in the final result, must vanish. Because the characteristic polynomial is made up of real coefficients, the roots of the polynomial always come in complex conjugate pairs. Thus, the summation over each of a complex root pair in the exponential (and other real valued functions) will give real final answer. Indeed, assuming that symbols a, b, c, d, g, h are real and computing the sum over a single complex conjugate root pair we come to the following relation,

$$\exp(a+ib)\frac{c+id}{g+ih}+\exp(a-ib)\frac{c-id}{g-ih}=\frac{2e^{a}\left((cg+dh)\cos b+(ch-dg)\sin b\right)}{g^{2}+h^{2}},$$

the right hand side of which formally represents a real number as expected. The left hand side is exactly the expression which we have in (6) and (9) formulas after summation over one pair of complex conjugate roots. However, from symbolic computation point of view the issue is not so simple. In general, the roots of high degree (when $d \geq 5$) polynomial equations cannot be solved in radicals and, therefore, in symbolic packages they are usually represented as the enumerated formal functions/algorithms of some irreducible polynomials. In *Mathematica* the formal solution is represented as **Root[poly, k]**. In order to obtain real answer, therefore, we have to know how to manipulate with these formal objects algebraically. To that end there exist algorithms which allow to rewrite the coefficients of irreducible polynomials **poly** after they have been algebraically manipulated. The operation, however, appears to be nontrivial and time consuming. In *Mathematica* it is implemented by **RootReduce**[] command, which produces another **Root[poly′**, k′] object. Such a root reduction typically raises the order of the irreducible polynomial. From pure numerical point of view, of course, we may safely remove spurious complex part in the final answer to get a real numerical value.

4 Elementary Functions of MV

Formulas (6) and (9) appear to be more universal than we have expected initially. In fact they allow to compute any function and transformation of MV (at least for diagonalizable MVs) if one replaces the exponential weight $\exp(\lambda_i)$ by any other function (and allows to use complex numbers). Here we shall demonstrate how to compute $\log(\mathsf{A}), \sinh(\mathsf{A})$, $\operatorname{arcsinh}(\mathsf{A})$ and Bessel $J_0(\mathsf{A})$ GA functions of MV A in $Cl_{4,0}$ that appeared in Example 2. The example with zero and negative eigenvalues was chosen to demonstrate that no problems arise if formal symbolic manipulations are addressed.

After replacement of $\exp(\lambda_i)$ by $\log(\lambda_i)$ in (9) and summing-up over all roots one obtains

$$\begin{aligned}\log \mathsf{A} =& \frac{1}{2}(\log(0_+) + \log(-8)) \\ &+ \frac{1}{8}(\log(-8) - \log(0_+))\left(\mathbf{e}_1 + \mathbf{e}_2 + \mathbf{e}_3 + \mathbf{e}_4 + 2\sqrt{3}\mathbf{e}_{1234}\right).\end{aligned} \tag{17}$$

We shall not attempt to explain what $\log(-8)$ means in $Cl_{4,0}$ since we want to avoid presence of complex numbers in real $Cl_{4,0}$. We shall assume, however, that $\exp\bigl(\log(-8)\bigr) = -8$ and $\exp\bigl(\log(0_+)\bigr) = \lim_{x\to 0_+} \exp\bigl(\log(x)\bigr) = 0$. Then it is easy to check that under these assumptions the exponentiation of $\log \mathsf{A}$ yields $\exp(\log(\mathsf{A})) = \mathsf{A}$, i.e., the log function in Eq. (17) is formal inverse of exp.

There are no problems when computing hyperbolic and trigonometric functions and their inverses[2]. Indeed, after replacing $\exp(\lambda_i)$ by $\sinh(\lambda_i)$, $\operatorname{arcsinh}(\lambda_i)$ and Bessel $J_0(\mathsf{A})$ in (9) one finds, respectively,

$$\begin{aligned}\sinh \mathsf{A} =& \frac{1}{8}\sinh(8)\bigl(-4 - \mathbf{e}_1 - \mathbf{e}_2 - \mathbf{e}_3 - \mathbf{e}_4 - 2\sqrt{3}\mathbf{e}_{1234}\bigr), \\ \operatorname{arcsinh} \mathsf{A} =& \frac{1}{8}\operatorname{arcsinh}(8)\bigl(-4 - \mathbf{e}_1 - \mathbf{e}_2 - \mathbf{e}_3 - \mathbf{e}_4 - 2\sqrt{3}\mathbf{e}_{1234}\bigr), \\ J_0(\mathsf{A}) =& \frac{1}{2}(1 + J_0(8)) + \frac{1}{8}(J_0(8) - 1)\bigl(\mathbf{e}_1 + \mathbf{e}_2 + \mathbf{e}_3 + \mathbf{e}_4 + 2\sqrt{3}\mathbf{e}_{1234}\bigr).\end{aligned} \tag{18}$$

It is easy to check that $\sinh\bigl(\operatorname{arcsinh}(\mathsf{A})\bigr) = \mathsf{A}$ is satisfied indeed. Here we do not question where special functions of the MV argument might be applied in practice. The purpose of the last procedure was aimed just to demonstrate that the formulas (6) and (9) allow to perform computations over a much larger class of functions and transformations related to MVs.

5 Conclusion

The paper shows that in Clifford geometric algebras the exponential of a general multivector is associated with the characteristic polynomial of the multivector and may be expressed in terms of roots of respective characteristic equation. In higher dimensional algebras the coefficients at basis elements, in agreement with [3], include a mixture of trigonometric and hyperbolic functions. The presented exponential formulas can be generalized to large class of trigonometric, hyperbolic functions and their inverses, as well as for fractional powers, special functions, etc.

A Minimal Polynomial of MV

A simple algorithm for computation of matrix minimal polynomial is given in [24]. It starts by constructing $d \times d$ matrix M and its powers $\{1, M, M^2, \ldots\}$ and

[2] It looks as if the complex numbers are inevitable in computing trigonometric functions in most of real Clifford algebras, except of $Cl_{3,0}$ as well as few others [5].

subsequently converting each of the matrices into vector of length $d \times d$. The algorithm then checks consequently the sublists $\{1\}$, $\{1, M\}$, $\{1, M, M^2\}$ etc. until the vectors in a running sublist are found to be linearly dependent. Once a linear dependence is established the algorithm returns a polynomial equation, in which the coefficients of linear combination are multiplied by proper powers of a chosen variable x.

In case of GA, the orthonormal basis elements $\mathbf{e}_J$ are linearly independent, therefore it is enough to construct vectors made from real coefficients of MV. Then, the algorithm starts searching when these vectors of coefficients become linearly dependent.

The vector constructed from MV matrix representation has $d^2 = \left(2^{\lceil \frac{n}{2} \rceil}\right)^2$ components. This coincides with a number of coefficients (2^n) in MV for Clifford algebras of even n and is twice less than a number of matrix elements $d \times d$ for odd n. The latter property can be easily understood if one remembers that for odd n the matrix representation of Clifford algebra has a block-diagonal form. Therefore, only a single block will suffice for required matrix algorithm. The Algorithm 1 below describes how to compute the minimal polynomial of MV without addressing to matrix representation.

Algorithm 1: Algorithm for finding minimal polynomial of MV in $Cl_{p,q}$

MinimalPoly(A)
Input: multivector $\mathsf{A} = a_0 + \sum_J^{2^n-1} a_J \mathbf{e}_J$ and polynomial variable x
Output: minimal polynomial $c_1 + c_2 x + c_3 x^2 + \cdots$
`/* Initialization */`
nullSpace={}; lastProduct=1; vectorList={};
`/* keep adding new MV coefficient vectors to vectorList until null space becomes nontrivial */`
While[nullSpace==={},
lastProduct=A∘lastProduct;
AppendTo[vectorList, ToCoefficientList[lastProduct]];
nullSpace=NullSpace[Transpose[vectorList]];
];
`/* use null space weights to construct the polynomial` $c_1 + c_2\mathsf{A} + c_3\mathsf{A}^2 + \cdots$`, with A replaced by given variable` x `*/`
return First[nullSpace] $\cdot \{x^0, x^1, x^2, \ldots, x^{\text{Length[nullSpace]}-1}\}$

All functions in the above code are internal *Mathematica* functions, except of ∘ (geometric product) and **ToCoefficientList**[ ] which is rather simple. The latter takes MV A and outputs a coefficient vector, i.e. **ToCoefficientList**[$a_0 + a_1\mathbf{e}_1 + a_2\mathbf{e}_2 + \cdots + a_I I$]→ $\{a_0, a_1, a_2, \ldots, a_I\}$. The real job is done by *Mathematica* function **NullSpace**[ ], which searches for linear dependency of inserted vector list. This function is a standard function of every linear algebra library. If the list of the vectors is linearly dependent it outputs weight factors of a linear combination for which the sum of vectors becomes zero, and an empty list otherwise. The **AppendTo[vectorList, newVector]** appends the **newVector** to the list of already checked vectors in **vectorList**.

References

1. Abdulkhaev, K., Shirokov, D.: On explicit formulas for characteristic polynomial coefficients in geometric algebras. In: Magnenat-Thalmann, N., et al. (eds.) CGI 2021. LNCS, vol. 13002, pp. 670–681. Springer, Cham (2021). https://doi.org/10.1007/978-3-030-89029-2_50
2. Acus, A., Dargys, A.: Square root of a multivector of Clifford algebras in 3D: a game with signs, pp. 1–29. arXiv:math-phi/2003.06873 (2020)
3. Acus, A., Dargys, A.: Coordinate-free exponentials of general multivector in Cl(p, q) algebras for p+q=3. Math. Meth. Appl. Sci. 1–13 (2022). https://doi.org/10.1002/mma.8529
4. Acus, A., Dargys, A.: Geometric algebra mathematica package. Tech. rep. (2022). https://github.com/ArturasAcus/GeometricAlgebra
5. Chappell, J.M., Iqbal, A., Gunn, L.J., Abbott, D.: Functions of multivector variables. PLoS ONE **10**(3), 1–21 (2015). https://doi.org/10.1371/journal.pone.0116943
6. Costa, V.R.: On the exponentials of some structured matrices. J. Phys. A **37**, 11613–11627 (2004)
7. Dargys, A., Acus, A.: Square root of a multivector in 3D Clifford algebras. Nonlinear Anal. Model. Control. **25**(3), 301–320 (2020)
8. Dargys, A., Acus, A.: Exponential of general multivector in 3D Clifford algebras. Nonlinear Anal. Model. Control. **27**(1) (2022). https://doi.org/10.15388/namc.2022.27.24476
9. Dargys, A., Acus, A.: Exponentials and logarithms of multivector in low dimensional (n=p+q¡3) Clifford algebras, pp. 1–14. arXiv:math-ph/2204.04895v1 (2022)
10. Dorst, L., Valkenburg, R.: Square root and logarithm of rotors in 3D conformal geometric algebra using polar decomposition, pp. 81–104. Springer, London (2011). https://doi.org/10.1007/978-0-85729-811-9_5
11. Fujii, K.: Exponentiation of certain matrices related to the four level system by use of the magic matrix. arXiv:math-ph/0508018v1 (2007)
12. Fujii, K., Oike, H.: How to calculate the exponential of matrices. Far East J. Math. Educ. **9**(1), 39–55 (2012)
13. Gürlebeck, K., Sprössig, W.: Quaternionic and Clifford Calculus for Physicists and Engineers. Wiley, Chichester (1998). ISBN: 978-0-471-96200-7
14. Hanson, A.J.: Visualizing Quaternions. Elsevier, Amsterdam (2006)
15. Helmstetter, J.: Characteristic polynomials in Clifford algebras and in more general algebras. Adv. Appl. Clifford Algebras **29**(2), 30 (2019). https://doi.org/10.1007/s00006-019-0944-5
16. Herzig, E., Ramakrishna, V., Dabkowski, M.K.: Note on reversion, rotation and exponentiation in dimensions five and six. J. Geom. Symmetry Phys. **35**, 61–101 (2014). https://doi.org/10.7546/jgsp-35-2014-61-101
17. Hitzer, E.: On factorization of multivectors in Cl(3,0), Cl(1,2) and Cl(0,3), by exponentials and idempotents. Complex Var. Elliptic Equ. 1–23 (2021). https://doi.org/10.1080/17476933.2021.2001462
18. Householder, A.S.: The Theory of Matrices in Numerical Analysis. Dover Publications Inc., New York (1975)
19. Lounesto, P.: Clifford Algebra and Spinors. Cambridge University Press, Cambridge (1997). ISBN-13: 978-0521599160
20. Marchuk, N.G., Shirokov, D.: Theory of Clifford Algebras and Spinors. Krasand, Moscow (2020). ISBN: 978-5-396-01014-7. (in Russian)

21. Moler, C., Loan, C.V.: Nineteen dubious ways to compute the exponential of a matrix, twenty-five years later. SIAM Rev. **45**(1), 3–49 (2003)
22. Ramakrishna, V., Zhou, H.: On the exponential of matrices in su(4). J. Phys. A **39**, 3021–3034 (2005). https://doi.org/10.1088/0305-4470/39/12/011
23. Roelfs, M., Keninck, S.D.: Graded symmetry groups: plane and simple, pp. 1–17. arXiv:math-phi/2107.03771v1 (2021)
24. Rowland, T., Weisstein, E.W.: Matrix minimal polynomial. From MathWorld-A Wolfram Web Resource (2022). https://mathworld.wolfram.com/MatrixMinimalPolynomial.html
25. Shirokov, D.S.: Classification of Lie algebras of specific type in complexified Clifford algebras. Linear Multilinear Algebra **66**(9), 1870–1887 (2018). https://doi.org/10.1080/03081087.2017.1376612
26. Shirokov, D.S.: On computing the determinant, other characteristic polynomial coefficients, and inverses in Clifford algebras of arbitrary dimension. Comput. Appl. Math. **40**(173), 1–29 (2021). https://doi.org/10.1007/s40314-021-01536-0
27. Hou, S.-H.: Classroom note: a simple proof of the Leverrier-Faddeev characteristic polynomial algorithm. SIAM Rev. **40**(3), 706–709 (1998)
28. Zela, F.D.: Closed-form expressions for the matrix exponential. Symmetry **6**, 329–344 (2014). https://doi.org/10.3390/sym6020329. ISSN 2073-8994

On Noncommutative Vieta Theorem in Geometric Algebras

Dmitry Shirokov[1,2](✉)

[1] HSE University, 101000 Moscow, Russia
dshirokov@hse.ru

[2] Institute for Information Transmission Problems of the Russian Academy of Sciences, 127051 Moscow, Russia
shirokov@iitp.ru

Abstract. In this paper, we discuss a generalization of Vieta theorem (Vieta's formulas) to the case of Clifford geometric algebras. We compare the generalized Vieta's formulas with the ordinary Vieta's formulas for characteristic polynomial containing eigenvalues. We discuss Gelfand – Retakh noncommutative Vieta theorem and use it for the case of geometric algebras of small dimensions. The results can be used in symbolic computation and various applications of geometric algebras in computer science, computer graphics, computer vision, physics, and engineering.

Keywords: Geometric algebra · Clifford algebra · Vieta theorem · noncommutative Vieta theorem · Vieta's formulas · characteristic polynomial

1 Introduction

In algebra, Vieta's formulas (or Vieta theorem) relate the coefficients of any polynomial to sums and products of its roots. These formulas are named after the famous French mathematician François Viète (or Franciscus Vieta). In this paper, we extend Vieta's formulas to geometric algebras. We discuss the noncommutative Vieta theorem in geometric algebras and compare it with the ordinary Vieta theorem.

In this paper, the notion of characteristic polynomial in geometric algebras is used. Note that the determinant is used to calculate the inverse in geometric algebras [3,13,17,18]. In [1,2,17], the explicit formulas for the characteristic polynomial coefficients $C_{(k)}$ are presented in the cases $n \leq 6$. The characteristic polynomial in geometric algebras is also discussed in [11] and used to solve the Sylvester and Lyapunov equations in [15,16].

In Sect. 2, we introduce generalized Vieta's formulas in geometric algebras and compare them with the ordinary Vieta's formulas for characteristic polynomial containing eigenvalues. The generalized Vieta's formulas do not contain eigenvalues. In Sect. 3, we discuss Gelfand – Retakh noncommutative Vieta

E. Hitzer et al. (Eds.): ENGAGE 2022, LNCS 13862, pp. 28–37, 2023.
https://doi.org/10.1007/978-3-031-30923-6_3

theorem for an arbitrary skew-field with some remarks. In Sect. 4, we apply noncommutative Vieta theorem to the geometric algebras $\mathcal{G}_{p,q}$ in the case of small dimensions $n = p + q$.

The results of this paper can be useful in symbolic computation and various applications of geometric algebras and characteristic polynomials in computer science, computer graphics, computer vision, physics, and engineering.

2 Generalized Vieta's Formulas in Geometric Algebras

Let us consider the real (Clifford) geometric algebra $\mathcal{G}_{p,q}$, $n = p+q \geq 1$ [5,12,14] with the generators e_a, $a = 1, 2, \ldots, n$ and the identity element $e \equiv 1$. The generators satisfy

$$e_a e_b + e_b e_a = 2\eta_{ab} e, \qquad a, b = 1, 2, \ldots, n,$$

where $\eta = (\eta_{ab})$ is the diagonal matrix with its first p entries equal to 1 and the last q entries equal to -1 on the diagonal. The grade involution and reversion of an arbitrary element (a multivector) $U \in \mathcal{G}_{p,q}$ are denoted by

$$\widehat{U} = \sum_{k=0}^{n} (-1)^k \langle U \rangle_k, \qquad \widetilde{U} = \sum_{k=0}^{n} (-1)^{\frac{k(k-1)}{2}} \langle U \rangle_k,$$

where $\langle U \rangle_k$ is the projection of U onto the subspace $\mathcal{G}^k_{p,q}$ of grade $k = 0, 1, \ldots, n$.

Let us consider the following faithful representation (isomorphism) of the complexified geometric algebra $\mathbb{C} \otimes \mathcal{G}_{p,q}$, $n = p + q$

$$\beta : \mathbb{C} \otimes \mathcal{G}_{p,q} \to M_{p,q} := \begin{cases} \mathrm{Mat}(2^{\frac{n}{2}}, \mathbb{C}) & \text{if } n \text{ is even}, \\ \mathrm{Mat}(2^{\frac{n-1}{2}}, \mathbb{C}) \oplus \mathrm{Mat}(2^{\frac{n-1}{2}}, \mathbb{C}) & \text{if } n \text{ is odd}. \end{cases} \quad (1)$$

The real geometric algebra $\mathcal{G}_{p,q}$ is isomorphic to some subalgebra of $M_{p,q}$, because $\mathcal{G}_{p,q} \subset \mathbb{C} \otimes \mathcal{G}_{p,q}$ and we can consider the representation of not minimal dimension

$$\beta : \mathcal{G}_{p,q} \to \beta(\mathcal{G}_{p,q}) \subset M_{p,q}.$$

We can introduce (see [17]) the notion of determinant

$$\mathrm{Det}(U) := \det(\beta(U)) \in \mathbb{R}, \qquad U \in \mathcal{G}_{p,q}$$

and the notion of characteristic polynomial

$$\varphi_U(\lambda) := \mathrm{Det}(\lambda e - U) = \lambda^N - C_{(1)}\lambda^{N-1} - \cdots - C_{(N-1)}\lambda - C_{(N)} \in \mathcal{G}^0_{p,q} \equiv \mathbb{R},$$
$$U \in \mathcal{G}_{p,q}, \qquad N = 2^{[\frac{n+1}{2}]}, \qquad C_{(k)} = C_{(k)}(U) \in \mathcal{G}^0_{p,q} \equiv \mathbb{R}, \quad k = 1, \ldots, N, \quad (2)$$

where $\mathcal{G}^0_{p,q}$ is a subspace of elements of grade 0, which we identify with scalars.

Let us denote the solutions of the characteristic equation $\varphi_U(\lambda) = 0$ (i.e. eigenvalues) by $\lambda_1, \ldots, \lambda_N$. By the Vieta's formulas from matrix theory, we know that

$$C_{(k)} = (-1)^{k+1} \sum_{1 \leq i_1 < i_2 < \cdots < i_k \leq n} \lambda_{i_1} \lambda_{i_2} \cdots \lambda_{i_k}, \qquad k = 1, \ldots, N,$$

in particular,

$$C_{(1)} = \lambda_1 + \cdots + \lambda_N = \mathrm{Tr}(U), \qquad \ldots, \qquad C_{(N)} = -\lambda_1 \cdots \lambda_N = -\mathrm{Det}(U),$$

where $\mathrm{Tr}(U) := \mathrm{tr}(\beta(U)) = N\langle U\rangle_0$ is the trace of U. The elements $C_{(k)}$, $k = 1, \ldots, N$ are elementary symmetrical polynomials in the variables $\lambda_1, \ldots, \lambda_N$.

2.1 The Case $n = 1$

Let us consider the particular case $n = 1$. In this case, the geometric algebra $\mathcal{G}_{p,q}$ is commutative and $N = 2$. We have

$$C_{(1)} = \lambda_1 + \lambda_2 \in \mathbb{R}, \qquad C_{(2)} = -\lambda_1\lambda_2 \in \mathbb{R}. \tag{3}$$

But also we have (see [17])

$$C_{(1)} = U + \widehat{U} \in \mathcal{G}^0_{p,q} \equiv \mathbb{R}, \qquad C_{(2)} = -U\widehat{U} \in \mathcal{G}^0_{p,q} \equiv \mathbb{R}. \tag{4}$$

The elements $y_1 := U$ and $y_2 := \widehat{U}$ are not scalars (and are not equal to the eigenvalues λ_1 and λ_2), but they are solutions of the characteristic equation $\varphi_U(x) = 0$, $x = y_1, y_2$ by the Cayley – Hamilton theorem (see the details in Sect. 4). Using

$$\lambda^2 - (U + \widehat{U})\lambda + U\widehat{U} = 0,$$

we get the explicit formulas for the eigenvalues

$$\begin{aligned} \lambda_{1,2} &= \frac{1}{2}(U + \widehat{U} \pm \sqrt{(U+\widehat{U})^2 - 4U\widehat{U}}) = \frac{1}{2}(U + \widehat{U} \pm \sqrt{(U-\widehat{U})^2}) \\ &= \langle U\rangle_0 \pm \sqrt{(\langle U\rangle_1)^2}, \end{aligned} \tag{5}$$

which do not coincide with the explicit formulas for $y_{1,2}$

$$y_{1,2} = \langle U\rangle_0 \pm \langle U\rangle_1, \tag{6}$$

Because the scalar $\sqrt{(\langle U\rangle_1)^2}$ does not coincide with the vector (element of grade 1) $\langle U\rangle_1$. We see that the role of the roots $\lambda_{1,2}$ (which are complex scalars) of the characteristic equation is played by some combinations $y_{1,2}$ of involutions of elements (which are not scalars). In the case of degenerate eigenvalues, we have $\langle U\rangle_1 = 0$ and the coincidence $\lambda_{1,2} = y_{1,2} = U = \langle U\rangle_0$.

2.2 The Case $n = 2$

Let us consider the particular case $n = 2$. We have $N = 2$ and

$$C_{(1)} = \lambda_1 + \lambda_2 \in \mathbb{R}, \qquad C_{(2)} = -\lambda_1\lambda_2 \in \mathbb{R}. \tag{7}$$

But also we have (see [17])

$$C_{(1)} = U + \widehat{\widetilde{U}} \in \mathcal{G}^0_{p,q} \equiv \mathbb{R}, \qquad C_{(2)} = -U\widehat{\widetilde{U}} \in \mathcal{G}^0_{p,q} \equiv \mathbb{R}. \tag{8}$$

Note that $U\widehat{\widetilde{U}} = \widehat{\widetilde{U}}U$ in the case $n = 2$. The elements $y_1 := U$ and $y_2 := \widehat{\widetilde{U}}$ are not scalars (and are not equal to the eigenvalues λ_1 and λ_2), but they are solutions of the characteristic equation $\varphi_U(x) = 0$, $x = y_1, y_2$ by the Cayley – Hamilton theorem (see the details in Sect. 4). Using

$$\lambda^2 - (U + \widehat{\widetilde{U}})\lambda + U\widehat{\widetilde{U}} = 0,$$

we get the explicit formulas for the eigenvalues

$$\begin{aligned}\lambda_{1,2} &= \frac{1}{2}(U + \widehat{\widetilde{U}} \pm \sqrt{(U + \widehat{\widetilde{U}})^2 - 4U\widehat{\widetilde{U}}}) = \frac{1}{2}(U + \widehat{\widetilde{U}} \pm \sqrt{(U - \widehat{\widetilde{U}})^2}) \\ &= \langle U\rangle_0 \pm \sqrt{(\langle U\rangle_1 + \langle U\rangle_2)^2},\end{aligned} \tag{9}$$

which do not coincide with the explicit formulas for $y_{1,2}$

$$y_{1,2} = \langle U\rangle_0 \pm (\langle U\rangle_1 + \langle U\rangle_2), \tag{10}$$

where the scalar $\sqrt{(\langle U\rangle_1 + \langle U\rangle_2)^2} = \sqrt{(\langle U\rangle_1)^2 + (\langle U\rangle_2)^2}$ is not equal to the expression $\langle U\rangle_1 + \langle U\rangle_2$. The role of the roots $\lambda_{1,2}$ (which are complex scalars) of the characteristic equation is played by some combinations $y_{1,2}$ of involutions of elements (which are not scalars).

In the case of degenerate eigenvalues, we have $\langle U\rangle_1 = \langle U\rangle_2 = 0$ and the coincidence $\lambda_{1,2} = y_{1,2} = U = \langle U\rangle_0$ in the case of two Jordan blocks; or $(\langle U\rangle_1)^2 = -(\langle U\rangle_2)^2 \neq 0$ and $\lambda_{1,2} = \langle U\rangle_0 \neq y_{1,2} = \langle U\rangle_0 \pm (\langle U\rangle_1 + \langle U\rangle_2)$ in the case of one Jordan block. For example, for $U = 5e + \frac{1}{2}(e_2 + e_{12})$, we have $\lambda_{1,2} = 5$ and $y_{1,2} = 5e \pm \frac{1}{2}(e_2 + e_{12})$ in the case $n = p = 2$, $q = 0$.

2.3 The Case $n = 3$

Let us consider the case $n = 3$. We have $N = 4$ and the formulas (see [17])

$$\begin{aligned}C_{(1)} &= U + \widehat{U} + \widetilde{U} + \widehat{\widetilde{U}}, \\ C_{(2)} &= -(U\widetilde{U} + U\widehat{U} + U\widehat{\widetilde{U}} + \widehat{U}\widehat{\widetilde{U}} + \widetilde{U}\widehat{\widetilde{U}} + \widehat{U}\widetilde{U}), \\ C_{(3)} &= U\widehat{U}\widetilde{U} + U\widehat{U}\widehat{\widetilde{U}} + U\widetilde{U}\widehat{\widetilde{U}} + \widehat{U}\widetilde{U}\widehat{\widetilde{U}}, \\ C_{(4)} &= -U\widehat{U}\widetilde{U}\widehat{\widetilde{U}}.\end{aligned} \tag{11}$$

These formulas look like the ordinary Vieta's formulas for eigenvalues:

$$\begin{aligned}C_{(1)} &= \lambda_1 + \lambda_2 + \lambda_3 + \lambda_4, \\ C_{(2)} &= -(\lambda_1\lambda_2 + \lambda_1\lambda_3 + \lambda_1\lambda_4 + \lambda_2\lambda_3 + \lambda_2\lambda_4 + \lambda_3\lambda_4), \\ C_{(3)} &= \lambda_1\lambda_2\lambda_3 + \lambda_1\lambda_2\lambda_4 + \lambda_1\lambda_3\lambda_4 + \lambda_2\lambda_3\lambda_4, \\ C_{(4)} &= -\lambda_1\lambda_2\lambda_3\lambda_4.\end{aligned} \tag{12}$$

The elements

$$y_1 := U = \langle U\rangle_0 + \langle U\rangle_1 + \langle U\rangle_2 + \langle U\rangle_3, \quad y_2 := \widetilde{U} = \langle U\rangle_0 + \langle U\rangle_1 - \langle U\rangle_2 - \langle U\rangle_3,$$
$$y_3 := \widehat{U} = \langle U\rangle_0 - \langle U\rangle_1 + \langle U\rangle_2 - \langle U\rangle_3, \quad y_4 := \widetilde{\widehat{U}} = \langle U\rangle_0 - \langle U\rangle_1 - \langle U\rangle_2 + \langle U\rangle_3,$$

are not scalars (and are not equal to the eigenvalues $\lambda_1, \lambda_2, \lambda_3, \lambda_4$), but they are solutions of the characteristic equation $\varphi_U(x) = 0$, $x = y_1, y_2, y_3, y_4$ by the Cayley – Hamilton theorem (see the details in Sect. 4).

We call the formulas (4), (8), (11) and their analogues for the cases $n \geq 4$ *generalized Vieta's formulas in geometric algebra.* The formulas (4), (8), (11) were proved in [17] using recursive formulas for the characteristic polynomial coefficients following from the Faddeev – LeVerrier algorithm. In this paper, we present an alternative proof of these formulas using the techniques of noncommutative symmetric functions (see Sects. 3 and 4).

3 On Gelfand – Retakh Noncommutative Vieta Theorem

Let us discuss the following Gelfand – Retakh theorem (known as the *noncommutative Vieta theorem* [10], see also [4,6]). In Sect. 4, we use it for the characteristic polynomial in geometric algebras.

Theorem 1 ([10]). *If $\{x_1, \ldots, x_N\}$ is an ordered generic set (i.e. Vandermonde quasideterminants v_k exist for all $k = 1, \ldots, N$) of solutions of the equation*

$$P_N(x) := x^N - a_1 x^{N-1} - \cdots - a_N = 0 \tag{13}$$

over a skew-field, then for $k = 1, 2, \ldots, N$:

$$a_k = (-1)^{k+1} \sum_{1 \leq i_1 < i_2 < \cdots < i_k \leq N} y_{i_k} \cdots y_{i_1},$$

where

$$y_k = v_k x_k v_k^{-1}.$$

In [10], the definition of Vandermonde quasideterminants v_k is given (see also [8,9]). In this paper, we use another definition of the elements v_k from [7]:

$$v_k = P_{k-1}(x_k) = x_k^{k-1} - (y_{k-1} + \cdots + y_1)x_k^{k-2} + \cdots + (-1)^{k-1} y_{k-1} \cdots y_1. \tag{14}$$

In particular, we have

$$v_1 = 1, \qquad v_2 = x_2 - y_1, \qquad v_3 = x_3^2 - (y_2 + y_1)x_3 + y_2 y_1. \tag{15}$$

Let us give examples.

In the case $N = 1$, substituting $x = x_1$ into (13), we obtain $y_1 = x_1 = a_1$ and $v_1 = 1$.

In the case $N = 2$, the Eq. (13) with $a_1 = y_2 + y_1$ and $a_2 = -y_2 y_1$ can be rewritten in the form

$$(x - y_1)x - y_2(x - y_1) = 0. \tag{16}$$

Substituting $x = x_1$ into (16), we obtain $(x_1 - y_1)x_1 - y_2(x_1 - y_1) = 0$ and we can take $y_1 = x_1$. Substituting $x = x_2$ into (16), we obtain

$$(x_2 - y_1)x_2 - y_2(x_2 - y_1) = 0 \tag{17}$$

and we can take $y_2 = v_2 x_2 v_2^{-1}$ in the case of invertible $v_2 := x_2 - y_1 = x_2 - x_1$. If $[x_2, v_2] = 0$, then $y_2 = x_2$ by (17).

In the case $N = 3$, the Eq. (13) with $a_1 = y_3 + y_2 + y_1$, $a_2 = -(y_3 y_2 + y_3 y_1 + y_2 y_1)$, and $a_3 = y_3 y_2 y_1$ can be rewritten in the form

$$(x^2 - (y_2 + y_1)x + y_2 y_1)x - y_3(x^2 - (y_2 + y_1)x + y_2 y_1) = 0. \tag{18}$$

Substituting $x = x_1$ and $x = x_2$ into (18), we conclude that we can take again $y_1 = x_1$ and $y_2 = v_2 x_2 v_2^{-1}$ in the case of invertible $v_2 = x_2 - y_1 = x_2 - x_1$. If $[x_2, v_2] = 0$, then $y_2 = x_2$. Substituting $x = x_3$ into (18), we obtain $(x_3^2 - (y_2 + y_1)x_3 + y_2 y_1)x_3 - y_3(x_3^2 - (y_2 + y_1)x_3 + y_2 y_1) = 0$ and we can take $y_3 = v_3 x_3 v_3^{-1}$ in the case of invertible $v_3 := x_3^2 - (y_2 + y_1)x_3 + y_2 y_1$. We get $y_3 = x_3$ in the case $[x_3, v_3] = 0$. And so on.

Remark 1. The condition $[v_k, x_k] = 0$ is equivalent to

$$[E_j, x_k] = 0, \qquad j = 1, \ldots, k-1,$$

where E_j, $j = 1, \ldots, k-1$ are noncommutative elementary symmetric polynomials in the variables $y_{k-1}, \ldots, y_1$:

$$E_1 = y_{k-1} + \cdots + y_1, \qquad E_{k-1} = y_{k-1} \cdots y_2 y_1.$$

For example, in the particular case $N = 4$, when all $[v_k, x_k] = 0$, $k = 1, \ldots, N$, we can take $y_k = x_k$, $k = 1, 2, 3, 4$, in the case

$$[x_2, x_1] = 0, \quad [x_3, x_2 x_1] = 0, \quad [x_3, x_2 + x_1] = 0, \quad [x_4, x_3 x_2 x_1] = 0,$$
$$[x_4, x_3 x_2 + x_3 x_1 + x_2 x_1] = 0, \quad [x_4, x_3 + x_2 + x_1] = 0.$$

We use this particular case below in $\mathcal{G}_{p,q}$ with $n = p + q = 3$.

4 Application of Noncommutative Vieta Theorem to Geometric Algebras

Let us apply Theorem 1 to the particular case of the characteristic polynomial $\varphi_U(\lambda)$ in geometric algebra $\mathcal{G}_{p,q}$. The elements $a_k = C_{(k)} \in \mathbb{R}$, $k = 1, \ldots, N$ from (13) are scalars now. We need N solutions $x_1, x_2, \ldots x_N$ of the characterstic equation $\varphi_U(x) = 0$. By the Cayley – Hamilton theorem, we can take $x_1 = U$:

$$\varphi_U(U) = 0. \tag{19}$$

We have the following statement.

Theorem 2. *We have*

$$\varphi_U(\lambda) = \varphi_{\widehat{U}}(\lambda) = \varphi_{\widetilde{U}}(\lambda) = \varphi_{\widehat{\widetilde{U}}}(\lambda), \tag{20}$$

$$\varphi_U(\widetilde{U}) = \varphi_U(\widehat{U}) = \varphi_U(\widehat{\widetilde{U}}) = 0. \tag{21}$$

Proof. We know that (see Lemma 10 in [17])

$$\mathrm{Det}(U) = \mathrm{Det}(\widehat{U}) = \mathrm{Det}(\widetilde{U}) = \mathrm{Det}(\widehat{\widetilde{U}}).$$

Using the definition of characteristic polynomial (2), we get

$$\varphi_{\widehat{U}}(\lambda) = \mathrm{Det}(\lambda e - \widehat{U}) = \mathrm{Det}(\widehat{\lambda e - U}) = \mathrm{Det}(\lambda e - U) = \varphi_U(\lambda).$$

Using the Cayley – Hamilton theorem $\varphi_U(U) = 0$, we get $\varphi_{\widehat{U}}(U) = 0$. Substituting $\widehat{U}$ for U, we get $\varphi_U(\widehat{U}) = 0$. We obtain the other formulas in a similar way. □

4.1 The Case $n = 1$

In this case, the geometric algebra is commutative. We can take $y_1 = x_1 = U$ in Theorem 1 by the Cayley – Hamilton theorem. The element $x_2 = \widehat{U}$ satisfies the characteristic equation (see Theorem 2). We have $v_2 = x_2 - x_1 = \widehat{U} - U = -2\langle U \rangle_1$. If $\langle U \rangle_1 \neq 0$, then $y_2 = x_2 = \widehat{U}$ and we obtain the formulas (4).

4.2 The Case $n = 2$

We can take $y_1 = x_1 = U$ in Theorem 1 by the Cayley – Hamilton theorem. The element $x_2 = \widetilde{U}$ satisfies the characteristic equation (see Theorem 2). We have $v_2 = x_2 - x_1 = \widetilde{U} - U = -2\langle U \rangle_2$. If $\langle U \rangle_2 = 0$, then we can use the formulas from the case $n = 1$. If $\langle U \rangle_2 \neq 0$, then $v_2 = \lambda e_{12}$, $\lambda \neq 0$ is invertible and $y_2 = v_2 \widetilde{U} v_2^{-1} = \widehat{\widetilde{U}}$ because the pseudoscalar e_{12} commutes with all even elements and anticommutes with all odd elements. We get the formulas (8).

4.3 The Case $n = 3$

We can take $y_1 = x_1 = U$ in Theorem 1 by the Cayley – Hamilton theorem.

Let us consider $x_2 = \widehat{\widetilde{U}}$. We have $v_2 = x_2 - x_1 = \widehat{\widetilde{U}} - U$ and $[v_2, x_2] = 0$ because $[U, \widehat{\widetilde{U}}] = 0$ in the case $n = 3$. Thus we can take $y_2 = x_2$.

Let us consider $x_3 = \widehat{U}$. We have

$$[x_3, v_3] = [x_3, x_3^2 - (x_1 + x_2)x_3 + x_2 x_1] = 0,$$

because the elements $x_1 + x_2 = U + \widehat{\widetilde{U}}$ and $x_2 x_1 = \widehat{\widetilde{U}} U$ belong to the center $\mathrm{Cen}(\mathcal{G}_{p,q}) = \mathcal{G}^0_{p,q} \oplus \mathcal{G}^3_{p,q}$, and can take $y_3 = x_3$.

Let us consider $x_4 = \widetilde{U}$. We have

$$[x_4, v_4] = [x_4, x_4^3 - (x_3 + x_2 + x_1)x_4^2 + (x_3x_2 + x_3x_1 + x_2x_1)x_4 - (x_3x_2x_1)] = 0,$$

because the elements $x_1 + x_2 = U + \widehat{\widetilde{U}}$ and $x_2x_1 = \widehat{\widetilde{U}}U$ belong to the center $\mathrm{Cen}(\mathcal{G}_{p,q})$, and $x_3x_4 = x_4x_3$, i.e. $\widehat{U}\widetilde{U} = \widetilde{U}\widehat{U}$ in the case $n = 3$. We take $y_4 = x_4$.

We obtain $y_k = x_k$, $k = 1, 2, 3, 4$ and the following formulas, which are another version of the formulas (11):

$$\begin{aligned}
C_{(1)} &= \widetilde{U} + \widehat{U} + \widehat{\widetilde{U}} + U,\\
C_{(2)} &= -(\widetilde{U}\widehat{U} + \widetilde{U}\widehat{\widetilde{U}} + \widetilde{U}U + \widehat{U}\widehat{\widetilde{U}} + \widehat{U}U + \widehat{\widetilde{U}}U),\\
C_{(3)} &= \widetilde{U}\widehat{U}\widehat{\widetilde{U}} + \widetilde{U}\widehat{U}U + \widetilde{U}\widehat{\widetilde{U}}U + \widehat{U}\widehat{\widetilde{U}}U,\\
C_{(4)} &= -\widetilde{U}\widehat{U}\widehat{\widetilde{U}}U.
\end{aligned} \tag{22}$$

Note that we obtain these formulas for the element U with invertible expressions v_2, v_3, and v_4 (for other elements U, other sequences x_1, x_2, x_3, x_4 can be considered). Also note that not every sequence y_1, y_2, y_3, y_4 from $\{\widetilde{U}, \widehat{U}, \widehat{\widetilde{U}}, U\}$ gives the correct Vieta's formulas (see Theorem 3 and Lemma 7 in [17], the formulas (11) and (22) are two of several correct forms).

4.4 The Cases $n \geq 4$

The generalized Vieta's formulas in the cases $n \geq 4$ are more complicated. We use the additional (triangle) operation (see [17])

$$U^{\triangle} := \sum_{k=0}^{n} (-1)^{\frac{k(k-1)(k-2)(k-3)}{24}} \langle U \rangle_k = \sum_{k=0,1,2,3 \mod 8} \langle U \rangle_k - \sum_{k=4,5,6,7 \mod 8} \langle U \rangle_k. \tag{23}$$

Note that

$$\mathrm{Det}(U^{\triangle}) \neq \mathrm{Det}(U), \qquad \varphi_{U^{\triangle}}(\lambda) \neq \varphi_U(\lambda), \qquad \varphi_U(U^{\triangle}) \neq 0 \tag{24}$$

in the general case (compare with the statements of Theorem 2).

In the case $n = 4$, the generalized Vieta's formulas have the following form

$$\begin{aligned}
C_{(1)} &= U + \widehat{\widetilde{U}} + \widehat{U}^{\triangle} + \widetilde{U}^{\triangle},\\
C_{(2)} &= -(U\widehat{\widetilde{U}} + U\widehat{U}^{\triangle} + U\widetilde{U}^{\triangle} + \widehat{\widetilde{U}}\widehat{U}^{\triangle} + \widehat{\widetilde{U}}\widetilde{U}^{\triangle} + (\widehat{U}\widetilde{U})^{\triangle}),\\
C_{(3)} &= U\widehat{\widetilde{U}}\widehat{U}^{\triangle} + U\widehat{\widetilde{U}}\widetilde{U}^{\triangle} + U(\widehat{U}\widetilde{U})^{\triangle} + \widehat{\widetilde{U}}(\widehat{U}\widetilde{U})^{\triangle},\\
C_{(4)} &= -U\widehat{\widetilde{U}}(\widehat{U}\widetilde{U})^{\triangle},
\end{aligned} \tag{25}$$

where the coefficients $C_{(k)}$, $k = 1, 2, 3, 4$ are not elementary symmetrical polynomials because of the additional operation of conjugation $\triangle$. These formulas look like the ordinary Vieta's formulas

$$
\begin{aligned}
C_{(1)} &= \lambda_1 + \lambda_2 + \lambda_3 + \lambda_4, \\
C_{(2)} &= -(\lambda_1\lambda_2 + \lambda_1\lambda_3 + \lambda_1\lambda_4 + \lambda_2\lambda_3 + \lambda_2\lambda_4 + \lambda_3\lambda_4), \\
C_{(3)} &= \lambda_1\lambda_2\lambda_3 + \lambda_1\lambda_2\lambda_4 + \lambda_1\lambda_3\lambda_4 + \lambda_2\lambda_3\lambda_4, \\
C_{(4)} &= -\lambda_1\lambda_2\lambda_3\lambda_4, \qquad (26)
\end{aligned}
$$

if we ignore the operation $\triangle$. The analogues of the formulas (25) for the cases $n = 5, 6$ are presented in [1] (see Theorem 5.1 and Sect. 8). These formulas also have the form of elementary symmetric polynomials, only if we ignore the operation $\triangle$, and can be interpreted as generalized noncommutative Vieta's formulas. These formulas do not follow directly from the Gelfand – Retakh noncommutative Vieta theorem, it is not easy task to guess the "right" (generic) ordered set of solutions $x_1, x_2, x_3, \ldots x_N$ of the characteristic equation to obtain the elements $y_1, y_2, y_3, \ldots, y_N$ we need in the generalized Vieta's formulas. This is a task for further research.

5 Conclusions

In this paper, we discuss a generalization of Vieta's formulas to the case of geometric algebras. We apply the Gelfand – Retakh theorem to the characteristic polynomial in geometric algebras. We show how to express characteristic coefficients in terms of combinations of various involutions of elements. We compare the generalized Vieta's formulas with the ordinary Vieta's formulas for eigenvalues. The role of the roots (which are complex scalars) of the characteristic equation is played by some combinations of involutions of elements (which are not scalars). The cases of small dimensions $n \leq 3$ are discussed in details. The case of arbitrary eigenvalues (including the case of degenerate eigenvalues) is considered. We plan to discuss Vieta's formulas in more complicated cases $n \geq 4$ in details using different techniques in the extended version of this paper. We also hope that the new approach presented in this paper (related to noncommutative symmetric functions) will help to find more optimized formulas for the determinant and inverse in geometric algebras in the cases $n \geq 6$.

Acknowledgements. This work is supported by the Russian Science Foundation (project 21-71-00043), https://rscf.ru/en/project/21-71-00043/.

The author is grateful to K. Abdulkhaev and N. Marchuk for useful discussions. The author is grateful to the four anonymous reviewers for their careful reading of the paper and helpful comments on how to improve the presentation.

References

1. Abdulkhaev, K., Shirokov, D.: Basis-free formulas for characteristic polynomial coefficients in geometric algebras. Adv. Appl. Clifford Algebras **32**, 57 (2022). https://link.springer.com/article/10.1007/s00006-022-01232-0

2. Abdulkhaev, K., Shirokov, D.: On explicit formulas for characteristic polynomial coefficients in geometric algebras. In: Magnenat-Thalmann, N., et al. (eds.) CGI 2021. LNCS, vol. 13002, pp. 670–681. Springer, Cham (2021). https://doi.org/10.1007/978-3-030-89029-2_50
3. Acus, A., Dargys, A.: The inverse of a Multivector: beyond the threshold $p+q=5$. Adv. Appl. Clifford Algebras **28**, 65 (2018)
4. Connes, A., Schwarz, A.: Matrix Vieta theorem revisited. Lett. Math. Phys. **39**, 349–353 (1997)
5. Doran, C., Lasenby, A.: Geometric Algebra for Physicists. Cambridge University Press, Cambridge (2003)
6. Fuchs, D., Schwarz, A.: Matrix Vieta Theorem, Amer. Math. Soc. Transl. ser. 2, vol. 169, Amer. Math. Soc., Providence (1995)
7. Fung, M.K.: On a simple derivation of the noncommutative Vieta theorem. Chin. J. Phys. **44**(5), 341–347 (2006)
8. Gelfand, I., Krob, D., Lascoux, A., Retakh, V., Thibon, J.-Y.: Noncommutative symmetric functions. Adv. Math. **112**, 218–348 (1995)
9. Gelfand, I., Retakh, V.: Quasideterminants. I, Selecta Math. **3**, 417–546 (1997)
10. Gelfand, I., Retakh, V.: Noncommutative Vieta theorem and symmetric functions. In: Gelfand, I.M., Lepowsky, J., Smirnov, M.M. (eds.) The Gelfand Mathematical Seminars, 1993–1995. Birkhäuser Boston. 1996. arxiv.org/abs/q-alg/9507010v1
11. Helmstetter, J.: Characteristic polynomials in Clifford algebras and in more general algebras. Adv. Appl. Clifford Algebras **29**, 30 (2019)
12. Hestenes, D., Sobczyk, G.: Clifford Algebra to Geometric Calculus - A Unified Language for Mathematical Physics. Reidel Publishing Company, Dordrecht Holland (1984)
13. Hitzer, E., Sangwine, S.: Multivector and multivector matrix inverses in real Clifford algebras. Appl. Math. Comput. **311**, 375–389 (2017)
14. Lounesto, P.: Clifford Algebras and Spinors. Cambridge University Press, Cambridge (1997)
15. Shirokov, D.: Basis-free solution to Sylvester equation in Clifford algebra of arbitrary dimension. Adv. Appl. Clifford Algebras **31**, 70 (2021)
16. Shirokov, D.: On basis-free solution to Sylvester equation in geometric algebra. In: Magnenat-Thalmann, N., et al. (eds.) CGI 2020. LNCS, vol. 12221, pp. 541–548. Springer, Cham (2020). https://doi.org/10.1007/978-3-030-61864-3_46
17. Shirokov, D.: On computing the determinant, other characteristic polynomial coefficients, and inverse in Clifford algebras of arbitrary dimension. Comput. Appl. Math. **40**, 173 (2021)
18. Shirokov, D.: Concepts of trace, determinant and inverse of Clifford algebra elements. In: Progress in analysis. Proceedings of the 8th congress of ISAAC, vol. 1, pp. 187–194. Peoples' Friendship University of Russia (2012). (ISBN 978-5-209-04582-3/hbk). arXiv:1108.5447

Transformations, Orientation and Fitting

Conjecture on Characterisation of Bijective 3D Digitized Reflections and Rotations

Stéphane Breuils[1(✉)], Yukiko Kenmochi[2], Eric Andres[3], and Akihiro Sugimoto[4]

[1] University of Savoie Mont-Blanc, LAMA Laboratory, Chambéry, France
stephane.breuils@univ-smb.fr
[2] Normandie Univ., UNICAEN, ENSICAEN, CNRS, GREYC, Caen, France
yukiko.kenmochi@unicaen.fr
[3] University of Poitiers, XLIM Laboratory, UMR CNRS 7252, Poitiers, France
eric.andres@univ-poitiers.fr
[4] National Institute of Informatics, Tokyo, Japan
sugimoto@nii.ac.jp

Abstract. Bijectivity of digitized linear transformations is crucial when transforming 2D/3D objects in computer graphics and computer vision. Although characterisation of bijective digitized rotations in 2D is well known, the extension to 3D is still an open problem. A certification algorithm exists that allows to verify that a digitized 3D rotation defined by a quaternion is bijective. In this paper, we use geometric algebra to represent a bijective digitized rotation as a pair of bijective digitized reflections. Visualization of bijective digitized reflections in 3D using geometric algebra leads to a conjectured characterization of 3D bijective digitized reflections and, thus, rotations. So far, any known quaternion that defines a bijective digitized rotation verifies the conjecture. An approximation method of any digitized reflection by a conjectured bijective one is also proposed.

1 Introduction

Geometric algebra has revealed its sufficient capability of handling linear transformations for geometric object manipulations, and has become a more powerful tool for the computer graphics and/or computer vision communities. In this paper, we propose to exploit *digitized linear transformations*, more specifically, digitized reflections and rotations, with the help of geometric algebra. The major problem with transformations in the digital world is that important properties may be lost. One of those crucial properties is bijectivity. Applying a transformation that is not bijective means that information may be simply lost or irreversibly altered (in case an interpolation is added in the process).

Bijective digitized rotations are a subject of study for almost thirty years now. First introduced in [1], the subset of angles for which digitized 2D rotations are bijective has been fully characterized [8,10,14]. Interesting links have

E. Hitzer et al. (Eds.): ENGAGE 2022, LNCS 13862, pp. 41–53, 2023.
https://doi.org/10.1007/978-3-031-30923-6_4

been made using Gaussian integers between twin Pythagorean triplets and the angles of digitized bijective rotations [14]. More recently, Pluta et al. [13] have brought a new light into this research subject by showing that similar results using Eisentein integers exist for the hexagonal grid.

In 3D and higher dimensions, characterization of bijective digitized rotations remains largely open. Pluta et al. [12] proposed a certification algorithm that confirms whether a given Lipschitz quaternion, which corresponds to a 3D rotation whose matrix representation is with only rational elements, defines a bijective digitized rotation, but characterization based on such quaternions remains elusive.

Meanwhile Andres et al. [2] proposed an algorithm for bijective digitized reflections in 2D which easily deduces a method that generates bijective digitized rotations. Breuils et al. [3] used geometric algebra to reformulate the problem and characterized 2D bijective digitized reflections. This is the starting point of the present paper. Here we look into the 3D characterization problem by using tools from geometric algebra [6,11] in order to overcome the problems of fundamentally handling 4D objects that Pluta et al. [12] encountered due to using quaternions. Geometric algebra allows us to establish a strong link between bijective digitized 3D reflections and bijective digitized 3D rotations. We first start by expressing the problem in geometric algebra's framework, and then focus more specifically on reflections, rotations, and the way in which Pluta et al. [12] described the problem with help of quaternions. This leads to a conjecture on characterization of rotation vectors corresponding to bijective digitized rotations in 3D, as well as the related bijective digitized reflections. The geometric algebra tools enable us to project Pluta et al. [12]'s quaternions into 3D and also visualize Pluta et al. [12]'s results, which so far match our proposed conjecture. The conjecture leads us to believe that all the cases where a digitized 3D rotation is bijective, correspond to 2D cases that are elevated to 3D. This greatly limits the scope of direct bijective digitized 3D rotations. If confirmed, it implies that further research will have to be conducted on approximated bijective digitized rotations. In the paper, we offer a different avenue to prove the conjecture that, if proven correct, would answer a thirty year old question. At the end of the paper, we propose an approximation method of any digitized reflection by a conjectured bijective one.

2 Digitized Reflections and Rotations via Geometric Algebra

Geometric algebra of a vector space is an algebra over a field such that the multiplication of vectors called the geometric product is defined on a space of elements, i.e., multivectors [6]. Geometric algebra is an intuitive and geometric object-oriented algebra that allows to define geometric transformations in an efficient way. Definitions and compositions of geometric transformations are given in the geometric algebra of $\mathbb{R}^3$, also called $\mathbb{G}_3$; see [6].

Let us briefly review reflections and rotations with geometric algebra. For more details, see [3,6,11]. We here focus on reflections and rotations expressed as two reflections. Since Pluta et al. [12] proposed a certification algorithm for

3D bijective digitized rotations represented by quaternions, we recall the link between quaternion algebra and geometric algebra. We then finish this section with the bijectivity condition and characterization of digitized reflections in 2D.

2.1 Reflections

A reflection is the isometric mapping from $\mathbb{R}^d$ to itself with a hyperplane as a set of fixed (invariant) points. It is defined as follows with geometric algebra when the hyperplane goes through the origin.

Definition 1. *Given a hyperplane passing through the origin, with its normal vector* $\mathbf{m} \in \mathbb{R}^d$*, denoted by* $H(\mathbf{m})$*, the reflection of point* $\mathbf{x} \in \mathbb{R}^n$ *with respect to* $H(\mathbf{m})$ *is defined as*

$$\left| \begin{array}{l} \mathcal{U}^{\mathbf{m}} : \mathbb{R}^d \to \mathbb{R}^d \\ \quad \mathbf{x} \mapsto -\mathbf{m}\mathbf{x}\mathbf{m}^{-1} = -\frac{1}{\|\mathbf{m}\|^2}\mathbf{m}\mathbf{x}\mathbf{m}. \end{array} \right.$$

Reflections $\mathcal{U}^{\mathbf{m}}$ are said rational if all the components of $\mathbf{m}$ are rational. Note that any rational reflection $\mathcal{U}^{\mathbf{m}}$ can be represented by $\mathbf{m} = \sum_{i=1\cdots d} u_i \mathbf{e}_i$ such that $u_i \in \mathbb{Z}$ and $\gcd(u_1, \cdots, u_d) = 1$.

2.2 Rotations

Any rotation is expressed as the composition of two reflections with geometric algebra. If a first reflection w.r.t. $H(\mathbf{m})$ followed by a second reflection w.r.t. $H(\mathbf{n})$, is applied to a point $\mathbf{x} \in \mathbb{R}^d$, we have the point $\mathbf{x}'$ such that

$$\mathbf{x}' = -\mathbf{n}(-\mathbf{m}\mathbf{x}\mathbf{m}^{-1})\mathbf{n}^{-1} = (\mathbf{n}\mathbf{m})\mathbf{x}(\mathbf{n}\mathbf{m})^{-1}. \tag{1}$$

In other words, $\mathbf{x}'$ is the rotation of $\mathbf{x}$ around the intersection of $\mathbf{m}$ and $\mathbf{n}$. Indeed, assuming $\mathbf{n}$ and $\mathbf{m}$ are both normalized, we have

$$\mathbf{x}' = (\cos\phi + \sin\phi\,\mathbf{I})\mathbf{x}(\cos\phi - \sin\phi\,\mathbf{I}), \tag{2}$$

where ϕ is the angle between $\mathbf{n}$ and $\mathbf{m}$ in the rotation plane whose bivector is $\mathbf{I}$. Note that the angle of this rotation corresponds to 2ϕ.

More generally, the algebraic entity representing the rotation of angle θ with respect to the rotation axis whose bivector is $\mathbf{U}$ is defined as

$$Q = \cos\frac{\theta}{2} + \sin\frac{\theta}{2}\frac{\mathbf{U}}{\|\mathbf{U}\|}. \tag{3}$$

Then, a point $\mathbf{x}$ is rotated to $\mathbf{x}'$ as follows:

$$\mathbf{x}' = Q\mathbf{x}Q^{\dagger}, \tag{4}$$

where $Q^{\dagger} = \cos\frac{\theta}{2} - \sin\frac{\theta}{2}\frac{\mathbf{U}}{\|\mathbf{U}\|}$.

2.3 Geometric Algebra Rotations and Quaternions

The subalgebra composed of the scalar and bivectors $\mathbb{R} \oplus \bigwedge^2 \mathbb{R}^3$ is isomorphic to the division ring of quaternions; see [9]. Let us consider a quaternion

$$q = a + b\mathbf{i} + c\mathbf{j} + d\mathbf{k}, \quad a, b, c, d \in \mathbb{R}, \tag{5}$$

where $\mathbf{i}^2 = \mathbf{j}^2 = -1$, $\mathbf{ij} = \mathbf{k}$ and $\mathbf{ki} = \mathbf{j}$, $\mathbf{jk} = \mathbf{i}$. The pure imaginary components can be related to the canonical basis of $\bigwedge^2 \mathbb{R}^3$ (bivectors) as follows:

$$\mathbf{i} = \mathbf{e}_{23}, \quad \mathbf{j} = \mathbf{e}_{13}, \quad \mathbf{k} = \mathbf{e}_{12}. \tag{6}$$

In $\mathbb{G}^3$, we can easily verify that $\mathbf{e}_{12}^2 = \mathbf{e}_{23}^2 = \mathbf{e}_{13}^2 = -1$ and

$$\mathbf{ij} = \mathbf{e}_{23}\mathbf{e}_{13} = \mathbf{e}_{12} = \mathbf{k}, \quad \mathbf{ki} = \mathbf{e}_{12}\mathbf{e}_{23} = \mathbf{e}_{13} = \mathbf{j}, \quad \mathbf{jk} = \mathbf{e}_{13}\mathbf{e}_{12} = \mathbf{e}_{23} = \mathbf{i}. \tag{7}$$

2.4 Cubic Grids and Cells and Digitized Reflections

In order to digitize points, we need a grid. In the following, we use the cubic grid also called the integer lattice defined as

$$\mathbb{Z}^d = \{\mathbf{x} = \sum_{i=1,\ldots,d} a_i \mathbf{e}_i \mid a_i \in \mathbb{Z}\}.$$

To a point κ on the cubic grid, it is handy to add the set of points that have the point κ as image after a rounding operation. This is called a digitized cell.

Definition 2 (transformed digitized cell). *Let us consider a transformation $\mathcal{T}$ such that any basis vector $\mathbf{e}_i$ is transformed to $\mathcal{T}\mathbf{e}_i\mathcal{T}^\dagger$. The digitization cell of $\kappa \in \mathbb{Z}^3$ transformed by $\mathcal{T}$ is defined as*

$$\begin{aligned} \mathcal{C}_\mathcal{T}(\kappa) := \{\mathbf{x} \in \mathbb{R}^d \mid \forall i \in [1, d] \;\; & \|\mathbf{x} - \kappa\| \leq \|\mathbf{x} - \kappa + \mathcal{T}\mathbf{e}_i\mathcal{T}^\dagger\| \\ \textit{and} \;\; & \|\mathbf{x} - \kappa\| < \|\mathbf{x} - \kappa - \mathcal{T}\mathbf{e}_i\mathcal{T}^\dagger\|\}. \end{aligned}$$

If $\mathcal{T} = 1$, i.e. $\mathcal{T}$ is the identity, $\mathcal{C}_1(\mathbf{0})$ is the typical digitized cell of the origin. Figure 1 shows elements of $\mathbb{Z}^3$ with a digitized cell associate to a point of the cubic grid $\mathbb{Z}^3$.

Definition 3 (Digitization operator). *The digitization operator on a cubic grid is defined as*

$$\left| \begin{array}{ccc} \mathcal{D}: & \mathbb{R}^d & \to & \mathbb{Z}^d \\ & \sum_{i=1,\ldots,d} u_i \mathbf{e}_i & \mapsto & \sum_{i=1,\ldots,d} \lfloor u_i + \frac{1}{2} \rfloor \mathbf{e}_i \end{array} \right. .$$

This allows to define a digitized reflection as the composition of a reflection and digitization.

Definition 4. *Given a hyperplane $H(\mathbf{m})$, a digitized reflection with respect to $H(\mathbf{m})$ is the composition of the reflection $\mathcal{U}^\mathbf{m}$ with the digitization operator $\mathcal{D}$ as follows*

$$\left| \begin{array}{ccc} \mathcal{R}^\mathbf{m}: \mathbb{Z}^d & \to & \mathbb{Z}^d \\ \mathbf{x} & \mapsto & \mathcal{D} \circ \mathcal{U}^\mathbf{m}(\mathbf{x}). \end{array} \right.$$

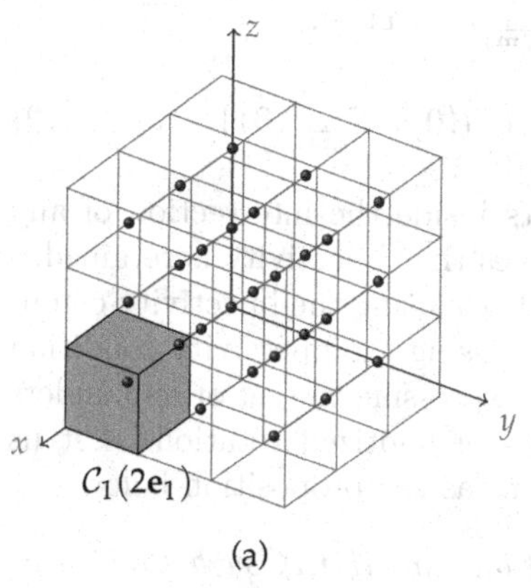

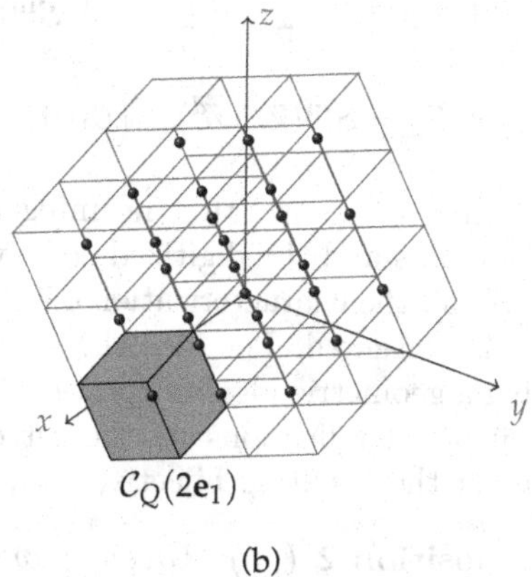

Fig. 1. (a) illustrates a set of points of $\mathbb{Z}^3$ as black points, whose associated digitization cells (voxel) are represented by wireframed cubes. The cube colored in blue is the digitization cell of $2\mathbf{e}_1$, i.e., $\mathcal{C}_1(2\mathbf{e}_1)$. (b) shows the rotated points and digitization cells by a geometric algebra rotation Q, where the cube colored in blue represents $\mathcal{C}_Q(2\mathbf{e}_1)$. (Color figure online)

2.5 Bijectivity Condition of Digitized Reflections and Characterization in 2D

In general, the digitization operator is not bijective, and, therefore, it is likely to produce holes and/or coincident points. However, there exist subsets of digitized transformations that are bijective. The characterization of these subsets were shown for digitized reflections in [3] and for rotations in [8,14].

For the characterization of bijective digitized reflections, the key idea is to investigate the structure of the set of remainders.

Definition 5. *Given a reflection $\mathcal{U}^{\mathbf{m}}$, the set of remainders $\mathcal{S}^{\mathbf{m}}$ is defined as*

$$\left| \begin{array}{rcl} \mathcal{S}^{\mathbf{m}} : \mathbb{Z}^d \times \mathbb{Z}^d & \to & \mathbb{R}^d \\ (\mathbf{x}, \mathbf{y}) & \mapsto & \mathcal{U}^{\mathbf{m}}(\mathbf{x}) - \mathbf{y}. \end{array} \right.$$

Given the set of remainders, the bijectivity condition is given as follows (see [14] also).

Proposition 1. *A digitized reflection $\mathcal{R}^{\mathbf{m}} = \mathcal{D} \circ \mathcal{U}^{\mathbf{m}}$ is bijective if and only if*

$$\forall \mathbf{y} \in \mathbb{Z}^d, \exists ! \mathbf{x} \in \mathbb{Z}^d, \mathcal{S}^{\mathbf{m}}(\mathbf{x}, \mathbf{y}) \in \mathcal{C}_1(\mathbf{0}),$$

where $\mathcal{C}_1(\mathbf{0})$ corresponds to origin-centered digitized cell.

Note that the above condition can be divided into two parts like [14]:

$$\begin{cases} \forall \mathbf{y} \in \mathbb{Z}^d, \exists \mathbf{x} \in \mathbb{Z}^d, \mathcal{S}^{\mathbf{m}}(\mathbf{x}, \mathbf{y}) \in \mathcal{C}_1(\mathbf{0}) \\ \forall \mathbf{x} \in \mathbb{Z}^d, \exists \mathbf{y} \in \mathbb{Z}^d, \mathcal{S}^{\mathbf{m}}(\mathbf{x}, \mathbf{y}) \in \mathcal{C}_{\frac{\mathbf{m}}{\|\mathbf{m}\|}}(\mathbf{0}) \end{cases}, \tag{8}$$

provided $\mathcal{S}^{\mathbf{m}}(\mathbb{Z}^d, \mathbb{Z}^d) \cap \mathcal{C}_1(\mathbf{0}) = \mathcal{S}^{\mathbf{m}}(\mathbb{Z}^d, \mathbb{Z}^d) \cap \mathcal{C}_{\frac{\mathbf{m}}{\|\mathbf{m}\|}}(\mathbf{0})$. Then,

$$\mathcal{I}_d = \mathcal{S}^{\mathbf{m}}(\mathbb{Z}^d, \mathbb{Z}^d) \cap \left(\mathcal{C}_1(\mathbf{0}) \cup \mathcal{C}_{\frac{\mathbf{m}}{\|\mathbf{m}\|}}(\mathbf{0})\right) \setminus \left(\mathcal{C}_1(\mathbf{0}) \cap \mathcal{C}_{\frac{\mathbf{m}}{\|\mathbf{m}\|}}(\mathbf{0})\right) = \emptyset. \tag{9}$$

Equation (9) shows that no integer point exists inside the intersection of any remainders and the digitized cells, which indicates that bijectivity is retained.

In [3], the characterization of digitized reflections using the bijectivity condition is presented. The idea there consists of expressing the bijectivity condition using a geometric algebra rotation $\mathbf{Q}$ in 2D and expressing the set of remainders of digitized reflections by the set of remainders of digitized rotations. Let us present the resulting bijective digitized reflections as the proposition below:

Proposition 2 ([3]). *Given a rational reflection line $H(\tilde{\mathbf{m}})$ such that $\tilde{\mathbf{m}} = -a\mathbf{e}_1 + b\mathbf{e}_2$, $a, b \in \mathbb{N}^*$, the rational digitized reflection $\mathcal{R}^{\tilde{\mathbf{m}}}$ is bijective if and only if $a = 1, b = 2k+1$ or $a = k, b = k+1$.*

3 Conjecture on the Characterization in 3D

We have seen that characterization of 2D bijective digitized reflections is known. In contrast, characterization of 3D bijective digitized reflections and rotations is an open problem. Pluta et al. [12] presented an algorithm that certifies whether or not a given Lipshitz quaternion (quaternion with integer components) is bijective. We start by making the same assumption as the conjecture of [12].

Conjecture 1 ([12]). *Given a vector $\mathbf{m} \in \mathbb{R}^d$, if one of the components of $\mathbf{m}$ is irrational, the digitized reflection with respect to the hyperplane $H(\mathbf{m})$ is not bijective.*

In order to give a conjecture on bijectivity, we first extend the certification algorithm [12] to digitized reflections. We then brute-force search bijective digitized reflections to capture an idea of their distributions. The brute-force search result yields a conjecture on 3D bijective digitized reflection. This conjecture enables us to deduce a conjecture on 3D bijective digitized rotations. Let us start by describing the certification algorithm.

3.1 Certification of Bijective Reflections Through Lipshitz Quaternions

The composition of a bijective digitized reflection (or non-bijective) with a bijective reflection is bijective (or non-bijective). Then, one possible certification algorithm for digitized reflections simply consists in composing the input normal vector with a reflection with respect to either the normal vector $\mathbf{e}_1, \mathbf{e}_2$ or $\mathbf{e}_3$. The result is a geometric algebra rotation and can be expressed with a Lipschitz quaternion [4]. Thus, the resulting geometric algebra rotation can be certified through Algorithm 1 of [12] with the four coefficients of the resulting geometric algebra rotation.

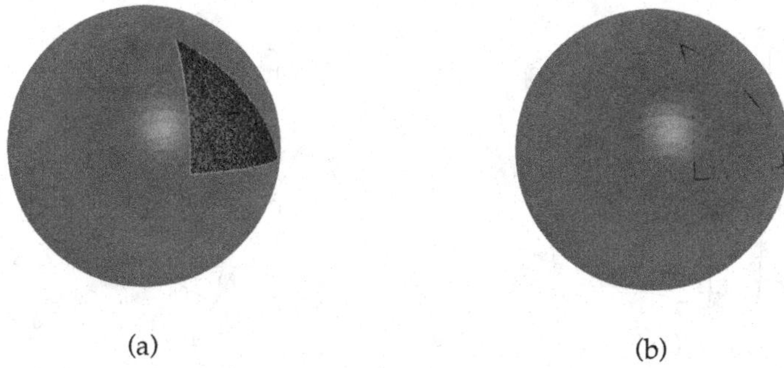

Fig. 2. (a) Red points are sampled unit normal vectors that are in Δ. Yellow circle arcs result from the boundary of Δ on the unit sphere (b) certified digitized reflections that are in Δ. (Color figure online)

Thanks to this algorithm, we can employ a brute-force search method for the bijective digitized reflections in a given window. Without loss of generality, let us study the bijectivity of digitized reflections in the domain Δ delimited as $\Delta = \{(x, y, z) \in \mathbb{Z}^3 \,|x \geq 0, y \geq 0, z \geq x + y\}$. Note that results in other domains can be obtained from octahedral symmetry of Δ.

The method of brute-force search for bijective digitized reflections in Δ is as follows. We start by sampling the domain Δ with normal vectors, and for each normal vector $\mathbf{m}_i$, we apply the geometric algebra certification algorithm as explained above. Both sampled $\mathbf{m}_i$ in Δ and the resulting certified transformations are shown in Fig. 2. Note that this result was obtained with the geometric algebra implementation ganja.js [5].

We observe in Fig. 2b that all the vectors $\mathbf{m}_i$ in Δ such that the digitized reflections $\mathcal{U}^{\mathbf{m}_i}$ are certified to be bijective are on the planes $\pi_1 : x = 0, \pi_2 : x = 0$ and $\pi_3 : z = x + y$. and there is no other digitized reflection $\mathcal{U}_{\mathbf{m}}$ such that $\mathbf{m}$ is outside the intersection of Δ and these planes. Furthermore, without loss of generality, given any conjectured bijective digitized reflection $\mathbf{m}_c \in \pi_1 \cap \Delta$, i.e., $\mathbf{m}_c = b\mathbf{e}_2 + c\mathbf{e}_3$ $(b, c \in \mathbb{N}, gcd(b, c) = 1)$, we find that either $b = k, c = k + 1$ $(k \in \mathbb{N})$ or $b = 1, c = 2k + 1$ $(k \in \mathbb{N})$. This latter observation suggests that any of the certified bijective digitized reflections can be expressed as an extension of the 2D bijective digitized reflections; see Fig. 5. This is the motivation of having the conjecture presented in the next section.

3.2 Bijective Digitized Reflections on Base Planes π_1, π_2, π_3

In this section, we focus on digitized reflections on the planes π_1, π_2, π_3 and give some conditions of bijectivity.

Proposition 3. *Any 3D digitized reflection $\mathcal{R}^{\mathbf{m}}$ such that $\mathbf{m} \in \pi_1 \cap \Delta$ is bijective iff*

$$\mathbf{m} = k\mathbf{e}_2 + (k + 1)\mathbf{e}_3 \quad or$$
$$\mathbf{m} = \mathbf{e}_2 + (2k + 1)\mathbf{e}_3,$$

where $k \in \mathbb{N}^*$.

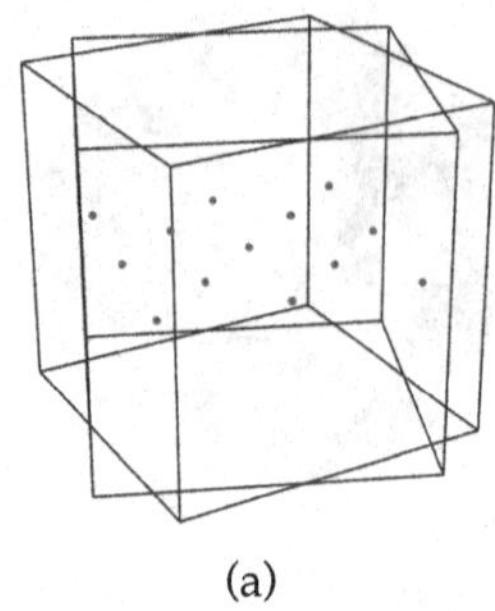
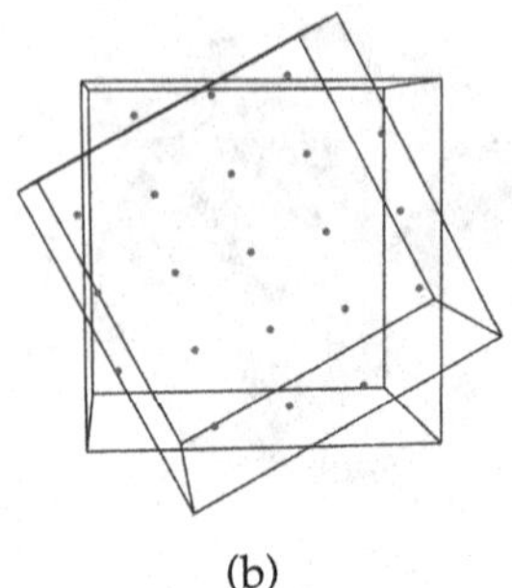

(a) (b)

Fig. 3. Dark green points are the elements of the set of remainders in $\mathcal{S}^{\mathbf{m}}(\mathbf{x}, \mathbf{y}) \cap \left(\mathcal{C}_1(\mathbf{0}) \cup \mathcal{C}_{\frac{\mathbf{m}}{\|\mathbf{m}\|}}(\mathbf{0})\right)$. Both $\mathcal{C}_1(\mathbf{0})$ and $\mathcal{C}_{\frac{\mathbf{m}}{\|\mathbf{m}\|}}(\mathbf{0})$ are denoted with blue cubes. (a) The digitized reflection is bijective since there is no element in $\mathcal{S}^{\mathbf{m}}(\mathbb{Z}^3, \mathbb{Z}^3) \cap \left(\mathcal{C}_1(\mathbf{0}) \cup \mathcal{C}_{\frac{\mathbf{m}}{\|\mathbf{m}\|}}(\mathbf{0})\right) \setminus \left(\mathcal{C}_1(\mathbf{0}) \cap \mathcal{C}_{\frac{\mathbf{m}}{\|\mathbf{m}\|}}(\mathbf{0})\right)$. (b) The digitized reflection is not bijective as there is one element of the set of remainders in each connected component of $\mathcal{S}^{\mathbf{m}}(\mathbb{Z}^3, \mathbb{Z}^3) \cap \left(\mathcal{C}_1(\mathbf{0}) \cup \mathcal{C}_{\frac{\mathbf{m}}{\|\mathbf{m}\|}}(\mathbf{0})\right) \setminus \left(\mathcal{C}_1(\mathbf{0}) \cap \mathcal{C}_{\frac{\mathbf{m}}{\|\mathbf{m}\|}}(\mathbf{0})\right)$. (Color figure online)

Proof. Let $\mathbf{m} = a\mathbf{e}_1 + b\mathbf{e}_2 + c\mathbf{e}_3$, then $a = 0$ and $c \geq b \geq 0$ as $\mathbf{m} \in \pi_1 \cap \Delta$. The set of remainders $\mathcal{S}^{\mathbf{m}}$, defined in Definition 5, is contained in the planes parallel to π_1, such as $x = n$ $(n \in \mathbb{Z})$; so the minimal distance between two planes of the set of remainders is 1. Furthermore, $\max(\mathcal{C}_1(\mathbf{0}) \cdot \mathbf{e}_1) = \max(\mathcal{C}_{\frac{\mathbf{m}}{\|\mathbf{m}\|}}(\mathbf{0}) \cdot \mathbf{e}_1) = 0.5$. Thus, the intersection $\mathcal{S}^{\mathbf{m}}(\mathbf{x}, \mathbf{y}) \cap \left(\mathcal{C}_1(\mathbf{0}) \cup \mathcal{C}_{\frac{\mathbf{m}}{\|\mathbf{m}\|}}(\mathbf{0})\right) \in \pi_1$. This yields

$$\mathcal{I}_3 = \mathcal{S}^{\mathbf{m}}(\mathbb{Z}^3, \mathbb{Z}^3) \cap \left(\mathcal{C}_1(\mathbf{0}) \cup \mathcal{C}_{\frac{\mathbf{m}}{\|\mathbf{m}\|}}(\mathbf{0})\right) \setminus \left(\mathcal{C}_1(\mathbf{0}) \cap \mathcal{C}_{\frac{\mathbf{m}}{\|\mathbf{m}\|}}(\mathbf{0})\right) \in \pi_1.$$

Therefore, the bijectivity condition can be rewritten and proved as $\mathcal{I}_2$ (Eq. (9)); Proposition 2 can be applied. □

Similarly, the following proposition holds as well.

Proposition 4. *Any 3D digitized reflection $\mathcal{R}^{\mathbf{m}}$ such that $\mathbf{m} \in \pi_2 \cap \Delta$ is bijective iff*

$$\mathbf{m} = k\mathbf{e}_1 + (k+1)\mathbf{e}_3 \quad \text{or}$$
$$\mathbf{m} = \mathbf{e}_1 + (2k+1)\mathbf{e}_3,$$

where $k \in \mathbb{N}$.

Examples of the set of remainders for these two last digitized reflections are shown in Fig. 3. Now let us consider the case where the reflection plane normal vectors $\mathbf{m}$ are on π_3.

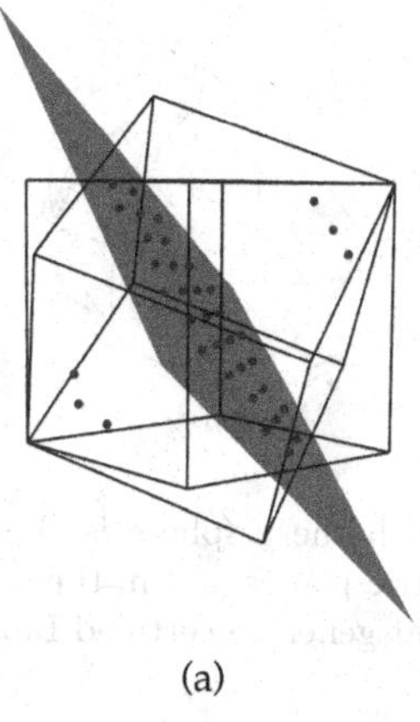

(a)

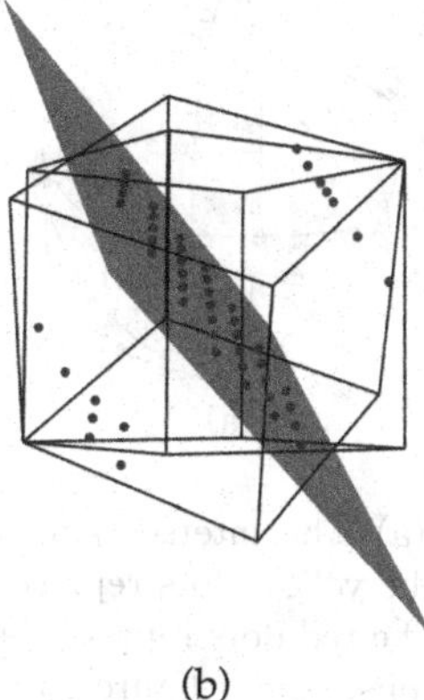

(b)

Fig. 4. Both $\mathcal{C}_1(\mathbf{0})$ and $\mathcal{C}_{\frac{\mathbf{m}}{\|\mathbf{m}\|}}(\mathbf{0})$ are denoted with a blue cube. The plane z=x+y is shown. Dark green points are the elements of the set of remainders in $\mathcal{S}^{\mathbf{m}}(\mathbf{x},\mathbf{y}) \cap \left(\mathcal{C}_1(\mathbf{0}) \cup \mathcal{C}_{\frac{\mathbf{m}}{\|\mathbf{m}\|}}(\mathbf{0})\right)$. Since no element of the set of remainders is in $\left(\mathcal{C}_1(\mathbf{0}) \cup \mathcal{C}_{\frac{\mathbf{m}}{\|\mathbf{m}\|}}(\mathbf{0})\right) \setminus \left(\mathcal{C}_1(\mathbf{0}) \cap \mathcal{C}_{\frac{\mathbf{m}}{\|\mathbf{m}\|}}(\mathbf{0})\right)$, the digitized reflection with respect to $\mathbf{m}$ is bijective (a), while it is not bijective (b) as some elements of the set of remainders are in $\left(\mathcal{C}_1(\mathbf{0}) \cup \mathcal{C}_{\frac{\mathbf{m}}{\|\mathbf{m}\|}}(\mathbf{0})\right) \setminus \left(\mathcal{C}_1(\mathbf{0}) \cap \mathcal{C}_{\frac{\mathbf{m}}{\|\mathbf{m}\|}}(\mathbf{0})\right)$. (Color figure online)

Proposition 5. *Any 3D digitized reflection* $\mathcal{R}^{\mathbf{m}}$ *such that* $\mathbf{m} \in \pi_3 \cap \Delta$ *is bijective iff*

$$\begin{aligned} \mathbf{m} &= k\mathbf{e}_1 + (k+1)\mathbf{e}_2 + (2k+1)\mathbf{e}_3 \quad \textit{or} \\ \mathbf{m} &= \mathbf{e}_1 + (2k+1)\mathbf{e}_2 + (2k+2)\mathbf{e}_3 \quad \textit{or} \\ \mathbf{m} &= (k+1)\mathbf{e}_1 + k\mathbf{e}_2 + (2k+1)\mathbf{e}_3 \quad \textit{or} \\ \mathbf{m} &= (2k+1)\mathbf{e}_1 + \mathbf{e}_2 + (2k+2)\mathbf{e}_3, \end{aligned}$$

where $k \in \mathbb{N}^*$.

Proof. Let $\mathbf{m} = a\mathbf{e}_1 + b\mathbf{e}_2 + c\mathbf{e}_3$, then $c \geq a, c = a + b$ as $\mathbf{m} \in \pi_3 \cap \Delta$. The set of remainders is on the planes parallel to π_3, for example, in $\mathcal{C}_1(\mathbf{0}) \cup \mathcal{C}_{\frac{\mathbf{m}}{\|\mathbf{m}\|}}(\mathbf{0})$:

$$\mathcal{S}^{\mathbf{m}}(\mathbf{x},\mathbf{y}) \cap \left(\mathcal{C}_1(\mathbf{0}) \cup \mathcal{C}_{\frac{\mathbf{m}}{\|\mathbf{m}\|}}(\mathbf{0})\right) \in \pi_3 \cup \pi_3 + 1 \cup \pi_3 - 1$$

The set of remainders for both bijective and non-bijective digitized reflections whose normal vectors are in $\pi_3 \cap \Delta$ is shown in Fig. 4. The set of remainders for both bijective and non-bijective digitized reflection whose normal vectors are in π_3 is shown in Fig. 4. These bijective or non-bijective normal vectors can be obtained from normal vectors of digitized reflections that are on the plane $z = 0$ through orthogonal projection. Furthermore, Propositions 3 and 4 extends well to digitized reflections whose normal vectors are on the projected plane. Thus, the bijectivity condition can be rewritten and proved as $\mathcal{I}_2$ (Eq. 9). □

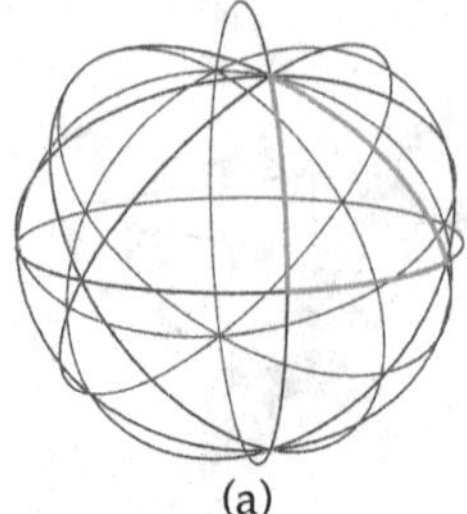

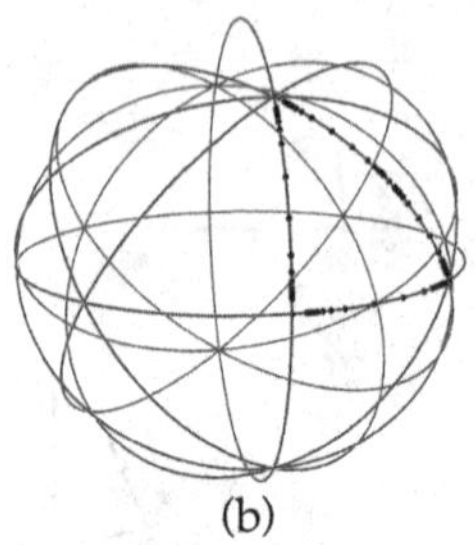

Fig. 5. (a) The intersection of the symmetric planes with the 2-sphere is shown in black. The yellow dots represent elements in the symmetric planes and in the domain Δ. (b) The red dots represent the bijective reflections that generate certified Lipschitz quaternions. (Color figure online)

3.3 Conjecture on Bijectivity for Digitized Reflections and Rotations

The three last propositions are the base for the following conjecture.

Conjecture 2. *The characterisation of 3D bijective digitized reflections is the extension of 2D bijective digitized reflections: a 3D digitized reflection is bijective if and only if it can be expressed with one of the normal vectors presented in Propositions 3, 4, and 5 or its octahedral symmetry.*

Any rotation is the product of two vectors in geometric algebra of $\mathbb{R}^3$. Moreover, it is possible to generalize the conjecture to all the 2-sphere using the octahedral symmetry and the planes π_1, π_2, π_3. The extension leads to all certified Lipschitz quaternion; this naturally allows to extend Conjecture 2.

Conjecture 3. *Any bijective digitized rotation in 3D can be defined as the composition of two (conjectured) bijective digitized reflections.*

4 Approximation with a Bijective Digitized Reflection

As seen in Fig. 2, the angular distribution of bijective digitized transformation is sparse. If our conjectures are valid, there would be a need to propose approximation methods for arbitrary angles. The idea of this section is to extend the approximation algorithm presented in [3] to $\mathbb{R}^3$ and approximate any digitized reflection with its nearest bijective one. First, let us consider the set of conjectured bijective digitized reflection as

$$\begin{aligned}\mathbf{B}_{k_{max}} = \{\mathcal{U}^{\tilde{\mathbf{m}}} \mid\ & \tilde{\mathbf{m}} = \lambda(1-\mu)k\mathbf{e}_1 + \mu(1-\lambda)k\mathbf{e}_2 + (k+1)\mathbf{e}_3,\\ & \tilde{\mathbf{m}} = \lambda(1-\mu)\mathbf{e}_1 + \mu(1-\lambda)\mathbf{e}_2 + (2k+1)\mathbf{e}_3,\\ & \tilde{\mathbf{m}} = (k+1)\mathbf{e}_1 + k\mathbf{e}_2 + (2k+1)\mathbf{e}_3,\\ & \tilde{\mathbf{m}} = 1\mathbf{e}_1 + (2k+1)\mathbf{e}_2 + (2k+2)\mathbf{e}_3,\\ & \lambda, \mu \in \{0,1\}, \lambda+\mu = 1, k \in \mathbb{N}^*, k \leq k_{max}\}.\end{aligned}$$

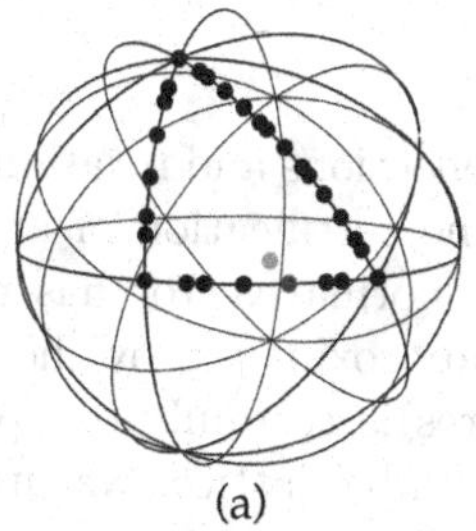
(a)

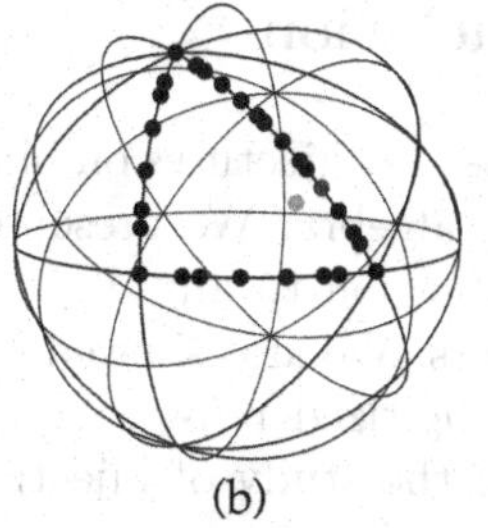
(b)

Fig. 6. Approximation of two digitized reflections, the normalized normal vector is shown with green dot, bijective normal vectors are shown with red dots and nearest bijective digitized reflection is shown in blue. The nearest symmetry plane is the $y = 0$ plane in (a) and $z = x + y$ plane (b). (Color figure online)

for a given $k_{max} \in \mathbb{N}^*$. We present here a straightforward way to approximate any digitized reflection $\mathcal{U}^{\mathbf{m}_i}$ with a bijective digitized reflection $\mathcal{U}^{\widetilde{\mathbf{m}}}$ of $\mathbf{B}_{k_{max}}$ for a given k_{max} such that

$$\arg\min_{\mathcal{U}^{\widetilde{\mathbf{m}}} \in \mathbf{B}_{k_{\max}}} d(\widetilde{\mathbf{m}}, \mathbf{m}_i) \tag{10}$$

where $d(\mathbf{a}, \mathbf{b})$ is the angular distance between two vectors $\mathbf{a}$ and $\mathbf{b}$.

For this, we look for $\widetilde{\mathbf{m}}$ that minimize the above objective on each plane of π_1, π_2, π_3. Let us consider that $\widetilde{\mathbf{m}} \in \pi_j$. Let $\mathbf{P}_j(\mathbf{m}_i)$ be the orthogonal projection of $\mathbf{m}_i$ into π_j. Then the above optimization on π_j consists in finding k_j that minimizes

$$\arg\min_{\mathcal{U}^{\widetilde{\mathbf{m}}} \in \mathbf{B}_{k_{\max}} \cap \pi_j} d(\widetilde{\mathbf{m}}, \mathbf{P}_j(\mathbf{m}_i)). \tag{11}$$

Without loss of generality, let us consider the case $j = 2$ with writing $\mathbf{P}_j(\mathbf{m}_i)) = (x, 0, z)$. Minimizing the above objective results in

$$\widetilde{k}_2 = \arg\min_{\widetilde{k} \in \{\lfloor \frac{x}{z-x} \rfloor, \lceil \frac{x}{z-x} \rceil, \lfloor \frac{z-x}{2x} \rfloor, \lceil \frac{z-x}{2x} \rceil\}} \left(\left| \frac{\widetilde{k}z - (\widetilde{k}+1)x}{(\widetilde{k}+1)z + \widetilde{k}x} \right|, \left| \frac{z - (2\widetilde{k}+1)x}{x + (2\widetilde{k}+1)z} \right| \right).$$

Note that this latter computation requires a constant time operation and does not depend on $k_{\max}$. Furthermore, the changes to perform for other symmetry planes merely consists in replacing x with y for the symmetry plane $x = 0$, and z with y for the symmetry plane $z = x + y$. The solution of (10) is the minimum among the three solutions of (11) for $j = 1, 2, 3$. Figure 6 shows the computation of the nearest bijective reflections of two non-bijective digitized reflections.

5 Conclusion

We proposed conjectures on bijective 3D digitized reflections and rotations using geometric algebra. We presented an extension of the certification of any Lipschitz quaternion to any digitized reflections whose normal vector has rational components. We also showed how any reflection is approximated by the nearest bijective digitized reflection. Proving the conjectures is certainly our perspective while the study of bijectivity is limited to the cubic lattice. Naturally, an extension of the presented conjectures to other 3D Bravais lattice with geometric algebra [7] is also expected as a perspective of this article. We are also interested in adapting the presented conjectures to the case where the number of points of $\mathbb{Z}^3$ is finite; in this case there would be more bijective digitized reflections and rotations.

References

1. Andres, E.: Cercles discrets et rotations discrétes. Ph.D. thesis, Université Louis Pasteur, Strasbourg, France (1992)
2. Andres, E., Dutt, M., Biswas, A., Largeteau-Skapin, G., Zrour, R.: Digital two-dimensional bijective reflection and associated rotation. In: Couprie, M., Cousty, J., Kenmochi, Y., Mustafa, N. (eds.) DGCI 2019. LNCS, vol. 11414, pp. 3–14. Springer, Cham (2019). https://doi.org/10.1007/978-3-030-14085-4_1
3. Breuils, S., Kenmochi, Y., Sugimoto, A.: Visiting bijective digitized reflections and rotations using geometric algebra. In: Lindblad, J., Malmberg, F., Sladoje, N. (eds.) DGMM 2021. LNCS, vol. 12708, pp. 242–254. Springer, Cham (2021). https://doi.org/10.1007/978-3-030-76657-3_17
4. Conway, J.H., Smith, D.A.: On Quaternions and Octonions: Their Geometry, Arithmetic, and Symmetry. AK Peters/CRC Press (2003)
5. De Keninck, S.: ganja.js (2020). https://doi.org/10.5281/ZENODO.3635774.https://zenodo.org/record/3635774
6. Dorst, L., Fontijne, D., Mann, S.: Geometric Algebra for Computer Science, An Object-Oriented Approach to Geometry. Morgan Kaufmann (2007)
7. Hestenes, D., Holt, J.W.: Crystallographic space groups in geometric algebra. J. Math. Phys. **48**(2), 023,514 (2007)
8. Jacob, M.A., Andres, E.: On discrete rotations. In: 5th International Workshop on Discrete Geometry for Computer Imagery, Clermont-Ferrand (France), pp. 161–174. Université de Clermont-Ferrand I (1995)
9. Lounesto, P.: Clifford algebras and spinors. In: Chisholm, J.S.R., Common, A.K. (eds.) Clifford Algebras and Their Applications in Mathematical Physics. ASIC, vol. 183, pp. 25–37. Springer, Dordrecht (1986). https://doi.org/10.1007/978-94-009-4728-3_2
10. Nouvel, B., Rémila, E.: Characterization of bijective discretized rotations. In: Klette, R., Žunić, J. (eds.) IWCIA 2004. LNCS, vol. 3322, pp. 248–259. Springer, Heidelberg (2004). https://doi.org/10.1007/978-3-540-30503-3_19
11. Perwass, C.: Geometric Algebra with Applications in Engineering, Geometry and Computing, vol. 4. Springer, Heidelberg (2009). https://doi.org/10.1007/978-3-540-89068-3

12. Pluta, K., Romon, P., Kenmochi, Y., Passat, N.: Bijectivity certification of 3D digitized rotations. In: Bac, A., Mari, J.-L. (eds.) CTIC 2016. LNCS, vol. 9667, pp. 30–41. Springer, Cham (2016). https://doi.org/10.1007/978-3-319-39441-1_4
13. Pluta, K., Roussillon, T., Cœurjolly, D., Romon, P., Kenmochi, Y., Ostromoukhov, V.: Characterization of bijective digitized rotations on the hexagonal grid. J. Math. Imaging Vis. **60**(5), 707–716 (2018)
14. Roussillon, T., Coeurjolly, D.: Characterization of bijective discretized rotations by Gaussian integers. Research report, LIRIS UMR CNRS 5205 (2016)

Complementary Orientations in Geometric Algebras

Leo Dorst(✉)

Computer Vision Group, University of Amsterdam, Amsterdam, The Netherlands
l.dorst@uva.nl

Abstract. Oriented elements are part of geometry, and they come in two complementary types: *intrinsic* and *extrinsic*. Those different orientation types manifest themselves by behaving differently under reflection. Dualization in geometric algebras can be used to encode them; or *vice versa*, orientation types inform the interpretation of dualization. We employ the Hodge dual, to include important algebras with null elements like PGA. Oriented elements can be combined using the meet operation, and the dual join (which is here introduced for that purpose). Software written to process one orientation type can be employed to process the complementary type consistently.

1 Oriented Geometry

In many applications of geometry, it makes sense to consider the geometric primitives as being oriented. In ray tracing, for instance, the rays are lines oriented in the direction of their propagation, and they interact with oriented surfaces which bound the interior of shapes. In mechanical motions, axes have a sense of turning – clockwise or counter-clockwise. There is thus a need for an algebra that computes consistently with such orientations.

However, many of the frameworks for geometric computation (such as 'homogeneous coordinates') are based in the mathematics of projective geometry, which traditionally neglects scalar factors independent of sign. Stolfi [1] was among the first to reintroduce signs in his thesis-based book 'Oriented Projective Geometry'. The rich illustrations of the various mathematical models, which were clearly meant to clarify, ultimately perhaps obscured the algebraic structure; the framework was never adopted universally.

In the last few decades, the formalization of geometric computations has a solid foundation in geometric algebra (or Clifford algebra) [2,3], with its ability to have not only subspaces as elements of computation (in a Grassmann-algebra manner), but also the ability to transform them universally with versors (representing orthogonal transformations), projection operators, etc. By choosing appropriate representation spaces with suitable metrics, this unifying framework can use versors to model changes of d-dimensional *attitude* (in the GA of directions $\mathbb{R}_d$), *Euclidean motions* (in the plane-based PGA $\mathbb{R}_{d,0,1}$ [4,5]), *conformal transformations* (in the conformal CGA $\mathbb{R}_{d+1,1}$ [2,6]), *3D projective transformations* (in the

E. Hitzer et al. (Eds.): ENGAGE 2022, LNCS 13862, pp. 54–66, 2023.
https://doi.org/10.1007/978-3-031-30923-6_5

3D line algebra $\mathbb{R}_{3,3}$ [7]), and more. However, also in geometric algebras the consistent processing of orientational aspects has been mostly neglected.

There are actually two complementary types of orientation we want to represent. A line in 3D, for instance, can be oriented 'along' (as in linear motion and momentum) or 'around' (using the line as an axis). We will call these an '*intrinsic*' and '*extrinsic*' orientation, respectively. Both are needed: when modelling 3D dynamics in PGA [4,8], the extrinsic line of an axis or rate is encoded as a 2-blade and the intrinsic line of a force or momentum as a dual 2-blade.

This paper aims to make orientation treatment integral to geometric algebras, by relating it directly to the common operation of (Hodge) dualization.

2 Two Complementary Types of Orientation

Let us consider some simple situations which we may want to represent in our algebraic framework, see Fig. 1. Actually doing so using current geometric algebras will convey a computational paradox in the next section.

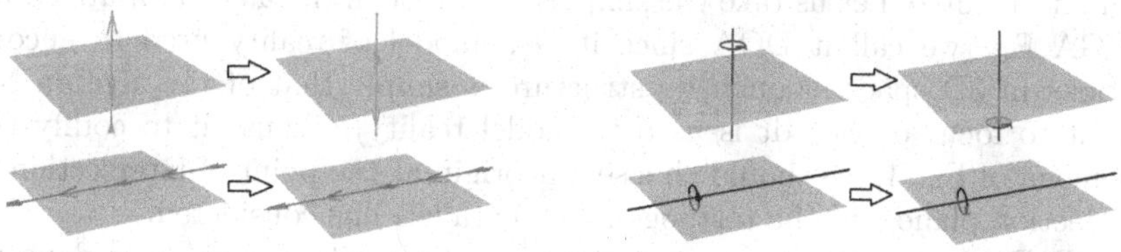

Fig. 1. Lines of complementary orientations reflect differently in a mirror.

In Fig. 1 top left, a plane is given, and an oriented line perpendicular to it. This line was made as the join of two points (not indicated), and therefore by the order of these points acquires the indicated orientation direction 'along' the line, which we will call *intrinsic orientation.* When we reflect the points in the indicated plane, the ordered line based on them acquires the opposite orientation. If we had instead considered an ordered line parallel to or even in the reflection plane (Fig. 1 bottom left), that line's orientation would have been preserved after reflection.

In Fig. 1 top right, we see an 'axis line' perpendicular to the reflection plane. Let us consider that extrinsically oriented line as having been constructed as the intersection of two orthogonal oriented planes (orthogonal to the reflection plane and to each other, not indicated). The two planes might have been oriented by labeling their 'front' and 'back'. Their meet line acquires an *extrinsic orientation* 'around' itself, by the ordering of the two planes whose meet it is (by the orientation of the shortest rotation angle that would achieve coincidence of the planes). This time, when we reflect the situation in the reflection plane, the two orthogonal planes remain unchanged (if extrinsic) or both swap signs (if intrinsic), and so their meet line after reflection is the same as before. By contrast, if we have a meet line parallel to the reflection plane (Fig. 1 bottom right), its orientation changes sign. It should, because when used as a rotation axis, the sense of rotation is reflected in the mirror.

That is how oriented geometry works in 3D. There are two types of lines: *join lines* (sometimes called 'spears') and *meet lines* (often called 'axes'), and they are both useful. For instance, when encoding 3D classical mechanics, join lines are local orbits or momenta of points, driven by motions that specified by meet line axes (finite for rotations, ideal for translations) [8]. The two types of lines show different reflection behavior, so they are geometrically and algebraically different.

Similarly, there are two types of planes: the ones with inside/out specification have an extrinsic orientation like , and the ones made as the join of points have an intrinsic orientation like . The former may be used to denote local inside/outside of a planar patch of a composite object, the latter for handedness-preserving local texture mapping.

3 A Representational Paradox

However, when we start encoding the situation in geometric algebra, we appear to run into trouble. Let us take the simplest GA model, the *algebra of directions* 3D DGA $\mathbb{R}_3$ (we call it DGA since it is a model of reality used to encode directions in 3D space; though its structure is simply that of the algebra $\mathbb{R}_3$; we want to focus on how it is used to model reality). To use it to codify the phenomena of Fig. 1, we should choose our origin at the point of intersection of the reflection plane and the (orthogonal or parallel) line considered.

In 3D DGA, reflection of a vector $\mathbf{x}$ in a plane with normal $\mathbf{n}$ is done by $\mathbf{x} \mapsto -\mathbf{n}\mathbf{x}\mathbf{n}^{-1}$ [2]. This is simply the conversion to geometric algebra of the familiar reflection formula $\mathbf{x} \mapsto \mathbf{x} - 2\frac{\mathbf{x}\cdot\mathbf{n}}{\mathbf{n}\cdot\mathbf{n}}\mathbf{n}$ of linear algebra, reflecting a vector $\mathbf{x}$ in an origin plane with normal vector $\mathbf{n}$ (converted to GA by replacing the dot product with the geometric product through $\mathbf{a}\cdot\mathbf{b} = (\mathbf{ab}+\mathbf{ba})/2$).

So we apparently use vectors in DGA to represent planes through the origin by their normal vector $\mathbf{n}$. Therefore in this simple model, it makes sense to consider the reflected $\mathbf{x}$ also to be the normal vector of a plane; that gives a single type of algebraic vector element which can be consistently interpreted geometrically in DGA. For multiple reflections by a versor V, made as the geometric product of a number of $|V|$ vectors, applied to an element X of grade $|X|$, the sandwiching formula becomes:

$$X \mapsto \underline{V}[X] \equiv (-1)^{|X||V|}\, V\, X\, V^{-1}. \tag{1}$$

The derivation is immediate from the vector reflection formula, see e.g. [3]. Although here briefly motivated by DGA, Eq. (1) is actually the general formulation in any GA. Only the geometric semantics of the algebraic vectors differs per model; the algebra is always the same. Moreover, any geometric algebra has a wedge product $\wedge$; it geometrically represents the intersection of planes, so we refer to it as the *meet*. It produces k-blades from vectors. There is also a join operation $\vee$, but we will be obliged to modify it; more about that later in Sect. 8.

The sign occurring in Eq. (1) is the natural sign one expects for multiple reflections of elements, in the normal representation of extrinsically characterized mirrors. Note that for even versors (the 'motions' of the algebra), there is no nett sign change: orientations can only swap under nett reflection.

To be concrete, let us provide 3D DGA with an orthonormal basis of planes $\{\mathbf{e}_1, \mathbf{e}_2, \mathbf{e}_3\}$. We take as mirror the plane $\mathbf{e}_3$. The two orthogonal planes $\mathbf{e}_1$ and $\mathbf{e}_2$ meet in the extrinsically oriented line $\mathbf{e}_1 \wedge \mathbf{e}_2 \equiv \mathbf{e}_{12}$ of Fig. 1(top right). If we reflect the situation, Eq. (1) gives $\mathbf{e}_1 \mapsto \underline{\mathbf{e}_3}[\mathbf{e}_1] = -\mathbf{e}_3\mathbf{e}_1\mathbf{e}_3 = \mathbf{e}_1$ and hence the plane $\mathbf{e}_1$ is preserved. The plane $\mathbf{e}_2$ is preserved as well, and therefore so is the meet line $\mathbf{e}_{12}$; we can of course also compute that preservation directly as $\underline{\mathbf{e}_3}[\mathbf{e}_{12}] = \mathbf{e}_3\, \mathbf{e}_{12}\, \mathbf{e}_3 = \mathbf{e}_{12}$. A meet line parallel to the mirror, such as $\mathbf{e}_{23}$ of Fig. 1(bottom right) reflects to $\mathbf{e}_{23} = \mathbf{e}_2 \wedge \mathbf{e}_3 \mapsto \mathbf{e}_2 \wedge (-\mathbf{e}_3) = -\mathbf{e}_{23}$, just as we motivated geometrically.

Now consider an intrinsically oriented line, geometrically coinciding with $\mathbf{e}_{12}$, see Fig. 1(top left). How are we going to represent it? The element $\mathbf{e}_{12}$ is no longer available, and moreover exhibits the wrong reflection behavior: we want a reflection in the $\mathbf{e}_3$-plane to change the sign of that intrinsic line in the $\mathbf{e}_3$-direction. The element $\mathbf{e}_3$ reflects properly $\mathbf{e}_3 \mapsto \underline{\mathbf{e}_3}[\mathbf{e}_3] = -\mathbf{e}_3\mathbf{e}_3\mathbf{e}_3 = -\mathbf{e}_3$, but in DGA it already has the meaning of (the normal vector of) a plane. So there is no element available, yet such a 'join line' is an oriented element that we would like to compute with. Somehow, we need to extend DGA to truly become the algebra of oriented directions, with both complementary types incorporated. DGA by itself does not seem big enough, since we have run out of algebraic elements.

4 Dualization

Like any geometric algebra, DGA has a dualization operation $\star$ (here geometrically corresponding to 'taking the orthogonal complement'). This dualization implies that an element of the algebra can be written in two ways, with possible consequences for its geometrical interpretation. For instance, the vector $\mathbf{e}_3$ represents the externally oriented plane; but algebraically, $\mathbf{e}_3 = \star\mathbf{e}_{12}$, so the element $\mathbf{e}_{12}$ could also be seen as characterizing geometry related to that plane. To consistently identify that geometrical interpretation, we should be guided by how algebraic elements and their duals transform.

We first derive how 'dual' elements reflect differently than 'direct' elements under a versor V. The difference is a sign that is related to the parity of the versor:

$$\underline{V}[\star X] = (-1)^{|V|} \star (\underline{V}[X]). \tag{2}$$

The proof should be straightforward once we have defined the dualization. Any linear dualization works in principle, but let us focus on the Hodge dual. The *Hodge dual* $\star$ is a linear operation, and therefore can be defined for arbitrary X by decomposing it on an orthogonal basis of blades $X = \sum_i X_i \mathbf{E}_i$, and specifying how dualization works on the orthogonal basis elements. (Here 'orthogonal basis' means that the basis blades are composed from a basis of orthogonal basis vectors.) We choose an implicit definition:

$$\textit{Hodge} \star \textit{on orthogonal basis element} : \mathbf{E}_i\,(\star\mathbf{E}_i) \equiv \mathcal{I}, \tag{3}$$

where $\mathcal{I}$ is the pseudoscalar of the algebra; a chosen sign for the pseudoscalar therefore fixes the sign of the Hodge dual. For d-D DGA $\mathbb{R}_d$, we use $\mathcal{I} = \mathbf{I}_d \equiv \mathbf{e}_1\mathbf{e}_2\cdots\mathbf{e}_d$; for d-D PGA $\mathbb{R}_{d,0,1}$, we use $\mathcal{I} = \mathbf{e}_0\mathbf{I}_d$. Since in a general algebra, $\mathbf{E}_i$ may be null, we could not simply define $\star\mathbf{E}_i \equiv \mathbf{E}_i^{-1}\mathcal{I}$ instead of Eq. (3).

Algebraically, Eq. (3) states that, for any orthogonal basis element $\mathbf{E}_i$ (like $\mathbf{e}_{01}$in 3D PGA), the dual contains 'the other indices' (like $\mathbf{e}_{23}$ in 3D PGA), and possibly a permutation sign. For a general $X = \sum_i X_i\mathbf{E}_i$ we simply have $\star X = \sum_i X_i \star \mathbf{E}_i$, since dualization is defined to be linear.

With that definition of the Hodge dual, we can attempt to prove Eq. (2) first for the basis elements, then for general X. From Eq. (1), a versor V applied as sandwich to the pseudoscalar gives $\underline{V}[\mathcal{I}] = (-1)^{|V|}\mathcal{I}$, the factor being the determinant of the map $\underline{V}[]$. We can write $\mathcal{I} = \mathbf{E}_i \star \mathbf{E}_i$ which transforms to $\underline{V}[\mathcal{I}] = \underline{V}[\mathbf{E}_i]\,\underline{V}[\star\mathbf{E}_i]$. But we also have $\mathcal{I} = \underline{V}[\mathbf{E}_i] \star \underline{V}[\mathbf{E}_i]$ by the definition of the Hodge dual for the element $\underline{V}[\mathbf{E}_i]$. We thus find $\underline{V}[\mathbf{E}_i] \star \underline{V}[\mathbf{E}_i] = (-1)^{|V|}\underline{V}[\mathbf{E}_i]\,\underline{V}[\star\mathbf{E}_i]$. If $\mathbf{E}_i$ is invertible, then so is $\underline{V}[\mathbf{E}_i]$ and we can deduce that $\underline{V}[\star\mathbf{E}_i] = (-1)^{|V|} \star \underline{V}[\mathbf{E}_i]$. By linearity, this then extends from $\mathbf{E}_i$ to general X, giving Eq. (2). □

Note that the last part of this derivation does not work for null elements, even though the Hodge dual is well-defined for them. Such null elements occur in very useful algebras like PGA, so we need to be able to handle them. Our solution is to *postulate* Eq. (2) for *all* intrinsic elements, as the map $\overline{\underline{V}}[]$ by which intrinsic elements should transform:

$$\overline{\underline{V}}[\star X] \equiv (-1)^{|V|} \star \underline{V}[X]. \tag{4}$$

The right hand side is always well defined, since the versor acts on a direct element X of the algebra.

With this equation for the transformation of Hodge duals, we can perform transformations on them without getting out of the algebra – there is thus no practical need to introduce a dual space to store the dual elements. You do need to know whether an element is to be considered as dual, though.

Applying Eq. (4) requires retrieving X from $\star X$. This is the *Hodge undualization*, which generally differs from dualization by a possible sign:

$$\star^{-1} X = (-1)^{(n-1)|X|} \star X, \tag{5}$$

where $|X|$ denotes the grade of X, and n the dimension of the representational space (so for d-D PGA $\mathbb{R}_{d,0,1}$ we have $n = d+1$). For an orthogonal-basis blade $\mathbf{E}_k$ of grade k: $(\star\mathbf{E}_k)(\star(\star\mathbf{E}_k)) = \mathcal{I} = \mathbf{E}_k(\star\mathbf{E}_k) = (-1)^{k(n-k)}(\star\mathbf{E}_k)\mathbf{E}_k$, so that $\star^{-1}\mathbf{E}_k = (-1)^{k(n-k)} \star \mathbf{E}_k = (-1)^{k(n-1)} \star \mathbf{E}_k$. Equation (5) for general X then follows by linearity. □

It is now possible to tie duality to the objective geometric property of orientation type, for we observe that the sign in Eq. (4), which depends on whether $\underline{V}[]$ is a nett reflection, is very reminiscent of the difference in behavior between intrinsic and extrinsic orientations of geometric primitives observed in Fig. 1.

The central idea of this paper is that *(Hodge) duals are encoding the complementary orientations within the existing structure of any geometric algebra.*

5 Complementary Orientation by Hodge Dualization

Let us illustrate the dual relationships by an example in 3D DGA: consider the reflection in $\mathbf{e}_3$, so that $V = \mathbf{e}_3$. The orthogonal plane $\mathbf{e}_3$ changes sign under this reflection in $\mathbf{e}_3$, since $\underline{\mathbf{e}_3}[\mathbf{e}_3] = -\mathbf{e}_3$, which is . The element $\star\mathbf{e}_{12}$ is algebraically equivalent to $\mathbf{e}_3$, since $\star\mathbf{e}_{12} = \mathbf{e}_3$. And indeed, $\star\mathbf{e}_{12}$ also changes sign: $\underline{\mathbf{e}_3}[\star\mathbf{e}_{12}] = \mathbf{e}_3(\star\mathbf{e}_{12})\mathbf{e}_3 = -\star(\mathbf{e}_3\mathbf{e}_{12}\mathbf{e}_3) = -\star\mathbf{e}_{12}$.

At this point, we may not quite know how to interpret $\star\mathbf{e}_{12}$, other than that it has a geometry related to the extrinsic line $\mathbf{e}_{12}$; should we view $\star\mathbf{e}_{12}$ as the intrinsic line , or as the orthogonal intrinsic plane ? But we have just seen that of the two choices to interpret $\star\mathbf{e}_{12}$, the intrinsic line changes sign under reflection in $\mathbf{e}_3$, while the geometry shows that is unchanged. This suggests that $\star\mathbf{e}_{12}$ should be interpreted as the intrinsic line . That is indeed confirmed by noting that $\mathbf{e}_3$ and $\star\mathbf{e}_{12}$ also behave identically under the other two reflections in the planes $\mathbf{e}_1$ and $\mathbf{e}_2$: they are both invariant.

Similarly, the dual rewriting of the algebraic element $\mathbf{e}_{12}$ is the equivalent $\star\mathbf{e}_3$, and this should be interpreted as the intrinsic plane : they both change sign under an $\mathbf{e}_1$-reflection or $\mathbf{e}_2$-reflection, and are invariant under $\mathbf{e}_3$-reflection.

We thus find that dual rewriting of algebraically equivalent elements produces geometrical objects with identical symmetries, and that they can be interpreted as being of the complementary type of orientation. More precisely, one algebraic element A can be parametrized in two different ways: *direct* as A, giving the extrinsic orientation as geometric interpretation (this was the implicit understanding in most earlier texts); and *dually* as $A = \star B$, giving the intrinsically oriented version of a geometric element B.

In Fig. 2 we suggest an iconic visualization of the various elements of 3D DGA and 3D PGA; this helps guide a more intuitive application of the algebras.

6 Visualization of Oriented 3D DGA Primitives

The algebra 3D DGA has extrinsic elements that are normal directions (of varying dimensions) at one location, which we will call the origin.

Planes: The extrinsically oriented coordinate planes are the basis elements of this algebra, and they are visualized as $\mathbf{e}_1$ and $\mathbf{e}_2$ and $\mathbf{e}_3$. Their duals, the intrinsically oriented planes, are $\star\mathbf{e}_1$ and $\star\mathbf{e}_2$ and $\star\mathbf{e}_3$.

Lines: The intersection of two coordinate planes produces the corresponding extrinsically oriented meet line. This gives $\mathbf{e}_{23}$ and $\mathbf{e}_{31}$ and $\mathbf{e}_{12}$. Their duals are the intrinsically oriented lines $\star\mathbf{e}_{23}$ and $\star\mathbf{e}_{31}$ and $\star\mathbf{e}_{12}$.

Point: 3D DGA has only one point, the meet $\mathbf{e}_{123}$ of the coordinate planes. It is extrinsically oriented, which we may indicate by a right-handed screw symbol like $\mathbf{e}_{123}$. Its dual $\star\mathbf{e}_{123}$ is that same origin point, but now with an intrinsic orientation. We denote it by $\star\mathbf{e}_{123}$, with open point and open arrowhead.

Volume: At the other end of the blade spectrum, there is the element 1. It represents the externally oriented volume element, which we tentatively denote by an un-anchored right hand screw symbol 1. Its dual is the intrinsically oriented volume element, which we could denote by $\star 1$, with open arrowhead.

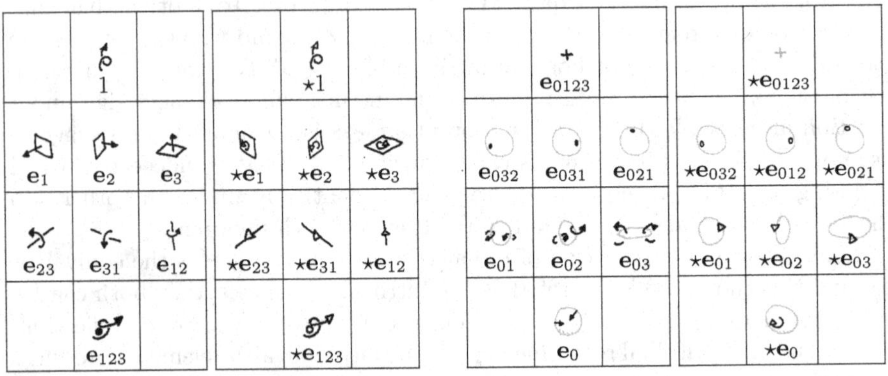

	1	
$\mathbf{e}_1$	$\mathbf{e}_2$	$\mathbf{e}_3$
$\mathbf{e}_{23}$	$\mathbf{e}_{31}$	$\mathbf{e}_{12}$
	$\mathbf{e}_{123}$	

	$\star 1$	
$\star\mathbf{e}_1$	$\star\mathbf{e}_2$	$\star\mathbf{e}_3$
$\star\mathbf{e}_{23}$	$\star\mathbf{e}_{31}$	$\star\mathbf{e}_{12}$
	$\star\mathbf{e}_{123}$	

	$\mathbf{e}_{0123}$	
$\mathbf{e}_{032}$	$\mathbf{e}_{031}$	$\mathbf{e}_{021}$
$\mathbf{e}_{01}$	$\mathbf{e}_{02}$	$\mathbf{e}_{03}$
	$\mathbf{e}_0$	

	$\star\mathbf{e}_{0123}$	
$\star\mathbf{e}_{032}$	$\star\mathbf{e}_{012}$	$\star\mathbf{e}_{021}$
$\star\mathbf{e}_{01}$	$\star\mathbf{e}_{02}$	$\star\mathbf{e}_{03}$
	$\star\mathbf{e}_0$	

Fig. 2. Left: The oriented basis elements of 3D DGA, the algebra of normal directions of origin planes. Right: The additional oriented elements for 3D PGA, the algebra of general planes. The gray sphere denotes the ideal plane at infinity.

7 Visualization of Oriented 3D PGA Primitives

Since Euclidean motions can be represented as multiple reflections in offset (hyper)planes, they form the versors of an algebra in which the vectors represent geometric hyperplanes. This is PGA (plane-based geometric algebra) $\mathbb{R}_{d,0,1}$ [4,5]. It parametrizes (hyper)planes through their normal vectors as $\mathbf{n} - \delta\mathbf{e}_0$, similar to $[\mathbf{n}^\top, \delta]$ in 'homogeneous coordinates'. The PGA embedding consistently includes Euclidean elements like offset lines and points, as meet and join of multiple planes.

Since 3D PGA subsumes 3D DGA, the total visualization table is Fig. 2.

Lines and Their Duals: In 3D PGA, lines and their duals are rather straightforward to depict. The element $\mathbf{e}_{12}$ is the extrinsically oriented line, in PGA as well as in DGA. It was equal to the dual $\star\mathbf{e}_3$ in DGA, the intrinsically oriented $\mathbf{e}_3$-plane. In PGA, $\mathbf{e}_{12}$ is equal to the dual $\star\mathbf{e}_{03}$ (since $\mathbf{e}_{03}\mathbf{e}_{12} = \mathbf{e}_{0123} = \mathcal{I}$), which is an ideal join line at infinity. We will depict the ideal elements in grey;

then the chosen icon of the intrinsically oriented ideal line $\star\mathbf{e}_{03}$ clearly encodes that its spatial oriented properties are similar to those of $\mathbf{e}_{12}$.

Conversely, in 3D PGA $\mathbf{e}_{03}$ is algebraically equal to $\star\mathbf{e}_{12}$. The ideal extrinsic line $\mathbf{e}_{03}$ is the 'ideal axis' of a translation in the $\mathbf{e}_3$-direction, and we visualize it as $\mathbf{e}_{03}$ (you can imagine how this 'axis' would act on a finite point in the middle by translating it upwards). The dual line $\star\mathbf{e}_{12}$ is intrinsic, and may be depicted as (just as it is in DGA). Both $\mathbf{e}_{03}$ and $\star\mathbf{e}_{12}$ clearly have the same reflection symmetries in the non-ideal coordinate planes $\mathbf{e}_1$, $\mathbf{e}_2$, $\mathbf{e}_3$. (There is no need to check reflection behavior in the ideal plane $\mathbf{e}_0$: since it is not invertible, it cannot reflect: so Eq. (1) is not defined for it.)

Planes and Points: In 3D PGA, the 1-vector (a geometrical plane) is dual to a 3-vector (a geometrical point). The exceptional ideal plane $\mathbf{e}_0$ has unusual reflection properties: it is invariant under the reflection in any coordinate plane $\mathbf{e}_1, \mathbf{e}_2, \cdots$, and we may depict it as (it has an inward extrinsic orientation, as you can see by taking the limit of an offset extrinsically oriented plane $\mathbf{n} - \delta\mathbf{e}_0$ for large δ). By contrast, its dual $\star\mathbf{e}_0$ equals $\mathbf{e}_{123}$, the oriented point at the origin. That acquires a minus sign under *any* reflection by Eq. (2). As in DGA, we depict it as $\mathbf{e}_{123}$, an origin point with a 3D screw-based orientation symbol.

The extrinsic coordinate plane $\mathbf{e}_3$ is visualized simply by its normal vector, and its dual $\star\mathbf{e}_3$ by the intrinsic orientation. This dual $\star\mathbf{e}_3$ is algebraically equal to $\star\mathbf{e}_3 = \mathbf{e}_{021}$, and the question arises how we show depict the latter element. Let us first see how it occurs geometrically.

Consider a point Z at location $\mathbf{z} = z\mathbf{e}_3$, which is represented as $Z = O + \mathbf{z}\mathcal{I} = \mathbf{e}_{123} + z\mathbf{e}_{021}$ (see [8]). As z becomes large this tends to become $\mathbf{e}_{021}$. Thus $\mathbf{e}_{021}$ can be conceived as the oriented point at infinity in the positive $\mathbf{e}_3$-direction, and we might depict it as , a solid point on the grey ideal plane. But under reflection in the $\mathbf{e}_3$-plane, $\underline{\mathbf{e}_3}[\mathbf{e}_{021}] = -\mathbf{e}_3\mathbf{e}_{021}\mathbf{e}_3 = \mathbf{e}_{021}$: it is invariant. That is not at all what we expect from the visualization $\mathbf{e}_{021}$ under an $\mathbf{e}_3$-reflection!

Algebraically, the point Z reflects to $\underline{\mathbf{e}_3}[Z] = -\mathbf{e}_{123} + z\mathbf{e}_{021} = -(\mathbf{e}_{123} - z\mathbf{e}_{021}) = -(O - \mathbf{z}\mathcal{I})$, which is a negatively oriented point at the reflected location $-\mathbf{z}$. From this, we understand why the direction $\mathbf{e}_{021}$ should indeed be invariant, or this double negation would not work. Thus indeed the infinite $\mathbf{e}_3$-direction should *not* change under $\mathbf{e}_3$-reflection; but the icon $\mathbf{e}_{021}$ does not convey this.

An algebraically more appropriate way of approaching $\mathbf{e}_{021}$ could be to write $\mathbf{e}_{021} = \mathbf{e}_{12} \wedge (-\mathbf{e}_0)$; this shows that $\mathbf{e}_{021}$ is the intersection of the extrinsically oriented line $\mathbf{e}_{12}$ with the invariant ideal plane $\mathbf{e}_0$, and that it thus inherits the symmetries of $\mathbf{e}_{12}$. A fairly faithful depiction of that construction would be $\mathbf{e}_{021}$, but this does not convey the point-like nature of the element $\mathbf{e}_{021}$.

It is a dilemma. We propose to use the icon $\mathbf{e}_{021}$, but with the understanding that its reflection properties are those of $\mathbf{e}_{12}$ (and hence of $\star\mathbf{e}_3$). Those are strange: under an $\mathbf{e}_3$-plane reflection, $\mathbf{e}_{021}$ must be invariant, but under an $\mathbf{e}_1$-reflection it should become its opposite $-\mathbf{e}_{021}$. So be it.

The dual $\star\mathbf{e}_{021}$ is an intrinsically oriented point at infinity. We depict it as , an open dot on the ideal plane. This actually has the same reflection symmetry as $\mathbf{e}_3$, and the intrinsic direction icon thus behaves as we would expect from a 1-dimensional direction element: in an $\mathbf{e}_3$-reflection it becomes . *Directions are intrinsically oriented ideal elements.*

Scalar and Pseudoscalar: The extreme elements of grade 1 and $d+1$ of the PGA algebra $\mathbb{R}_{d,0,1}$ are less obvious (or helpful?) to visualize.

The geometric positively oriented d-volume represented by the algebraic element 1 could be denoted by an unlocalized right-handed screw symbol, like in 3D, an unanchored version of the icon for $\mathbf{e}_{123}$, as in DGA.

The geometric dimension of the pseudoscalar $\mathbf{e}_{0123}$ is one less than that of a point. Such a geometric dimension of -1 is merely a scalar, so it could simply be denoted by a sign $_+$ for $\mathbf{e}_{0123}$ and $_-$ for $-\mathbf{e}_{0123}$.

The dual volume $\star\mathbf{e}_{0123}$ is algebraically identical to the element 1. We have as yet no clear intuition of how an intrinsic signed number $\star\mathbf{e}_{0123}$ should differ from its extrinsic counterpart, so for now denote it in grey as '$_+$'.

8 The Dual Join

From PGA (hyper)planes as basic extrinsically oriented elements, we can construct extrinsically oriented (hyper)lines, etc. by intersection. That geometric intersection operation (actually, piece-wise linear intersection, see [3]) is performed algebraically by the fundamental *meet* operation in any geometric algebra, the extension of the anti-symmetric (Grassmann) product $\wedge$ on the hyperplane vectors. When its arguments X and Y are transformed by a versor V, the meet product transforms like an 'outermorphism':

$$\underline{V}[X] \wedge \underline{V}[Y] = \underline{V}[X \wedge Y]. \tag{6}$$

As we have seen, dual elements will play a role in our geometric modelling, so we can wonder what the meet of duals of X and Y is. It may seem to make sense to express the outcome again as a dual (see e.g. [3]). We then define this total combination operation on X and Y as the *join*, denoted by $X \vee Y$:

$$\text{join:}\ \star(X \vee Y) \equiv \star X \wedge \star Y. \tag{7}$$

However, when we look at a specific result, we find for the join of two points (see [8]): $\mathbf{e}_{123} \vee (\mathbf{e}_{123} + \mathbf{e}_1\mathbf{e}_{0123}) = \mathbf{e}_{123} \vee \mathbf{e}_{032} = \star^{-1}(\mathbf{e}_0 \wedge \mathbf{e}_1) = \star^{-1}\mathbf{e}_{01} = \mathbf{e}_{23}$. Thus the join line connecting the two points separated in the positive $\mathbf{e}_1$-direction is represented as an *extrinsically* oriented element $\mathbf{e}_{23}$. In the context of oriented geometry, we would much rather have the result of joining two points in an *intrinsic* form, oriented intrinsically from the first point to the second, since that has the correct transformation symmetries. As we have seen, this is the dual $\star\mathbf{e}_{23}$ of the extrinsic result. So for the purposes of oriented geometry, we would rather define a new join $\triangledown$, which we could dub the *dual join*, through:

$$\text{dual join:}\ X \triangledown Y \equiv \star X \wedge \star Y. \tag{8}$$

Now the result of this dual join of the two points above is $\star\mathbf{e}_{23}$, the desired intrinsically oriented element. Note that the dual join is indeed the dual of the join: $X \triangledown Y = \star(X \vee Y)$, hence its suggested name.

The dual join transforms like an outermorphism under versors:

$$\underline{V}[X \triangledown Y] = \underline{V}[\star X \wedge \star Y] = \underline{V}[\star X] \wedge \underline{V}[\star Y] = \cdots = \underline{V}[X] \triangledown \underline{V}[Y]. \tag{9}$$

By contrast, the classical join of Eq. (7) contains an odd number of dualizations, and thus transforms as $\underline{V}[X \vee Y] = (-1)^{|V|}\, \underline{V}[X] \vee \underline{V}[Y]$ (see [5]).

The classical join is associative, but the dual join is somewhat awkward in this respect. It is not associative in the explicit form given for two arguments: you cannot simply apply that formula twice to compute $X \triangledown Y \triangledown Z$ as $(X \triangledown Y) \triangledown Z$. Rather, the extension of the dual join to more than two arguments should be done via its duality to the associative classical join:

$$X \triangledown Y \triangledown Z \equiv \star(X \vee Y \vee Z) = \star X \wedge \star Y \wedge \star Z.$$

Example: The PGA points $Q_0 = O = \mathbf{e}_{123}$, $Q_1 = O + \mathbf{e}_1\mathcal{I} = \mathbf{e}_{123} + \mathbf{e}_{032}$ and $Q_2 = O + \mathbf{e}_2\mathcal{I} = \mathbf{e}_{123} + \mathbf{e}_{013}$ classically join (see [5]) to form an extrinsic plane: $Q_0 \vee Q_1 \vee Q_2 = \star^{-1}(\star Q_0 \wedge \star Q_1 \wedge \star Q_2) = \star^{-1}(\star\mathbf{e}_{123} \wedge \star\mathbf{e}_{032} \wedge \star\mathbf{e}_{013}) = \star^{-1}(\mathbf{e}_{021}) = \mathbf{e}_3$. By contrast, the dual join $Q_0 \triangledown Q_1 \triangledown Q_2 = \star Q_0 \wedge \star Q_1 \wedge \star Q_2 = \star\mathbf{e}_3$ is the desired intrinsic plane $\star\mathbf{e}_3$.

We adopt the *meet* and the *dual join* as our preferred combination operations on oriented geometric elements. The meet preserves the orientation type, the dual join switches to the complementary type.

9 The Four Sibling Relationships

We now have two complementary ways of looking at oriented elements (intrinsic and extrinsic), and two dually related combination operations (meet and dual join) to produce new elements from a pair of existing oriented elements, be they intrinsic or extrinsic. These 2×2 possibilities combine to produce four closely related structural connections in oriented geometric algebra. To be specific, we take the equation $\mathbf{e}_3 \wedge \mathbf{e}_1 = \mathbf{e}_{31}$ and relate it to four associated expressions, by dualization of arguments and dualization of the combination operator.

Meeting Extrinsics: In the normal vector interpretation, the equation $\mathbf{e}_3 \wedge \mathbf{e}_1 = \mathbf{e}_{31}$ states how a line is made from the meet of two planes:

$$\mathbf{e}_3 \wedge \mathbf{e}_1 = \mathbf{e}_{31} \;\leftrightarrow\; \wedge \;=\; . \tag{10}$$

The resulting bivector $\mathbf{e}_{31}$ can be used to construct a rotation versor around the extrinsic axis $\mathbf{e}_{31}$ by exponentiation as $\exp(-\mathbf{e}_{31}\phi/2)$.

Meeting Intrinsics: Duality $\star$ in 3D DGA allows us to rewrite $\mathbf{e}_3 = \star\mathbf{e}_{12}$, $\mathbf{e}_1 = \star\mathbf{e}_{23}$ and $\mathbf{e}_{31} = \star\mathbf{e}_2$, so Eq. (10) can also be read as:

$$\star\,\mathbf{e}_{12} \wedge \star\mathbf{e}_{23} = \star\mathbf{e}_2 \tag{11}$$

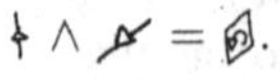

As the sketch shows that this retrieves the usual intuitive construction in the algebra of direction vectors (sometimes called VGA), with the usual outer product, to make an oriented bivector from the outer product of two vector directions.

Dual-Joining Extrinsics: We apply the dual join operation to rewrite Eq. (11:

$$\mathbf{e}_{12} \vee \mathbf{e}_{23} = \star\mathbf{e}_2 \tag{12}$$

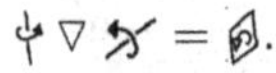

This denotes how two extrinsically oriented lines can be dual-joined to become an intrinsic plane, with an orientation determined by the smallest rotation.

Dual-Joining Intrinsics: Finally, we rewrite Eq. (12) in dual form:

$$\star\,\mathbf{e}_3 \vee \star\,\mathbf{e}_1 = \mathbf{e}_{31} \tag{13}$$

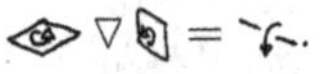

This shows how two intrinsic planes dual-join to form an extrinsic line, with the orientation again determined naturally by the smallest rotation.

The combination of *different* orientation types in 3D DGA mostly yields trivial equalities like $\mathbf{e}_3 \wedge \star\mathbf{e}_3 = \mathbf{e}_{123} = \star 1$ and $\star\mathbf{e}_3 \wedge \mathbf{e}_1 = 0$. We may investigate the geometric significance of such mixed combinations later, but not in this paper.

10 Computing with Complementary Orientations

The regular way of computing with PGA is based in the extrinsic 'planes as vectors' paradigm, which can naturally construct Euclidean motion versors as the exponentials of extrinsic bivectors (the 'axes'). In previous texts on PGA, that extrinsic orientation was assumed, often implicitly (though [8] briefly mentions the issue). The same holds for the other algebras: some orientation is implicitly assumed, often only to be gathered from how an author defines the sandwiching action for a single vector (the sign in their equivalent of our Eq. (1)).

We would of course also like to transform those new intrinsically oriented elements of the form $\star B$, since they are useful in modelling reality. This is simple, since Eq. (4) tells us how to revert to a corresponding transformation on an extrinsic element B:

$$\overline{\underline{V}}[\star B] = (-1)^{|V|} \star \underline{V}[B]. \tag{14}$$

There is therefore no real need to develop separate software to implement the intrinsic elements, including endowing them with their own sandwich product. But beware: using Eq. (14) is *not* the same as evaluating the extrinsic element A algebraically equivalent to $\star B$, transforming that, and rewriting in dual form!

As a PGA example, using the reflector $p = \mathbf{e}_3 - \delta\mathbf{e}_0$ (the plane offset by δ in the positive $\mathbf{e}_3$-direction), on the extrinsic plane $\mathbf{e}_3$ gives: $\mathbf{e}_3 \mapsto \underline{p}[\mathbf{e}_3] = -(\mathbf{e}_3 - \delta\mathbf{e}_0)\,\mathbf{e}_3\,(\mathbf{e}_3 - \delta\mathbf{e}_0) = -\mathbf{e}_3 + 2\delta\mathbf{e}_0 = -(\mathbf{e}_3 - 2\delta\mathbf{e}_0)$, which is the opposite plane $-\mathbf{e}_3$ offset by $2\delta\mathbf{e}_3$ from the origin. For the intrinsic $\star\mathbf{e}_3$ we find by Eq. (14): $\star\mathbf{e}_3 \mapsto \overline{p}[\star\mathbf{e}_3] = -\star\underline{p}[\mathbf{e}_3] = \star(\mathbf{e}_3 - 2\delta\mathbf{e}_0)$, the intrinsic plane offset by $2\delta\mathbf{e}_3$. However, had we naïvely evaluated $\star\mathbf{e}_3 = \mathbf{e}_{021}$, transformed that and rewritten in dual form, we would have found: $\underline{p}[\mathbf{e}_{021}] = -(\mathbf{e}_3 - \delta\mathbf{e}_0)\,\mathbf{e}_{021}(\mathbf{e}_3 - \delta\mathbf{e}_0) = -\mathbf{e}_{021} = -\star\mathbf{e}_3$; which is clearly wrong: the null elements wreak havoc and even the orientation sign is incorrect.

11 Conclusion

We have found that a single copy of a geometric algebra can support the representation of both extrinsic and intrinsically oriented elements. Those two complementary orientation types are related by dualization, and using the Hodge dual allows us to represent the duals within the original algebra. Since intrinsic elements transform differently than extrinsic elements under reflections, we do need to tag whether an element is intended to be used dually; if so, the correct reflection follows Eq. (4), so it may take an extra minus sign relative to the standard extrinsic sandwiching Eq. (1). For even versors, there is no sign difference between the complementary orientation types, which are both just moved along.

We used the Hodge dual throughout, since we have found algebras with null elements and a null pseudoscalar useful in applications; notably the plane-based PGA encoding Euclidean motions. The fundamental structure that supports complementary orientations can however be carried by any form of dual; we may soon consider employing the Poincaré dual [9] instead.

We found that elements of the same orientation type can be combined to produce an element of the same type by the *meet* operation, or of the complementary type by the *dual join*. That dual join is exactly dual to the join usually introduced in treatments of geometric algebras that ignore orientation types.

By courtesy of the (Hodge) dual, complementary orientation types can be accommodated in existing geometric algebras. This makes them compl*i*mentary as well as compl*e*mentary; so you might as well use them!

Acknowledgement. My sincere thanks to Steven De Keninck for catching a major oversight in the computational mapping, leading to refinement of Eq. (4).

References

1. Stolfi, J.: Oriented Projective Geometry, A Framework for Geometric Computations. Academic Press (2014)
2. Hestenes, D., Sobczyk, G.: Clifford Algebra to Geometric Calculus. Reidel (1984)
3. Dorst, L., Fontijne, D., Mann, S.: Geometric Algebra for Computer Science: An Object-oriented Approach to Geometry. Morgan Kaufman (2009)
4. Gunn, C.: Geometry, kinematics, and rigid body mechanics in Cayley-Klein geometries. Ph.D. dissertation, TU Berlin (2011)

5. Dorst, L., De Keninck, S.: Guided tour to the plane-based geometric algebra PGA (version 2.0) (2022). https://bivector.net/PGA4CS.html
6. Anglès, P.: Construction de revêtements du groupe conforme d'un espace vectorial muni d'une “métrique” de type (p, q). Annales de l'Institut Henri Poincaré, vol. Section A, vol. XXXIII, pp. 33–51 (1980)
7. Dorst, L.: 3D oriented projective geometry through versors of $\mathbb{R}^{3,3}$. Adv. Appl. Clifford Algebras **26**, 1137–1172 (2016)
8. Dorst, L., De Keninck, S.: May the Forque be with you, dynamics in PGA (version 2.2) (2022). https://bivector.net/PGADYN.html
9. Gunn, C.: A bit better: variants of duality in geometric algebras with degenerate metrics (2022). http://arxiv.org/2206.02459

On Proper and Improper Points in Geometric Algebra for Conics and Conic Fitting Through Given Waypoints

Pavel Loučka(✉)

Brno University of Technology, Technická 2896/2, 616 69 Brno, Czech Republic
Pavel.Loucka@vutbr.cz

Abstract. We introduce a new conic fitting algorithm using Geometric Algebra for Conics (GAC) that not only minimises the overall distance to a data set, but also causes the fitted conic to pass through prescribed waypoints. Moreover, the expression for points at infinity (also called improper points) in terms of GAC is derived; hence, the use of the improper waypoints in the conic fitting problem is enabled. Finally, a MATLAB implementation of the fitting algorithm and experimental results based on custom data sets are included.

Keywords: conic fitting · geometric algebra · Clifford algebra · proper point · point at infinity · ideal point · improper point · waypoint

1 Introduction

Geometric Algebra for Conics (GAC), originally introduced in [9], and consequently elaborated in [4], has already proven to be a useful tool for conic manipulation, [1], and for elementary conic fitting, [5], as well as conic fitting with additional geometric constraints such as axial alignment or location of the conic's centre at the origin of the coordinate system, [7,8].

However, other additional geometric constraints imposed on a fitted conic can be thought of. For example, we may demand that a conic fitted among data points should also pass through a given set of waypoints. Such a fit can be helpful, e.g. when computing the conical trajectories of dynamical systems numerically; at the beginning, some points of a trajectory are found and, afterwards, the conic is fitted among them, as in [2]. Nevertheless, the conic fitted by basic algorithms cannot ensure the fulfillment of the initial condition, i.e. that the resulting conic will pass through the corresponding initial point. As will be shown in Sect. 4, conic fitting through given waypoints can be performed using GAC in a way similar to the original GAC fitting algorithm described in [5] and briefly recalled in Sect. 2.

The research was supported by a BUT grant no. FSI-S-23-8161.

E. Hitzer et al. (Eds.): ENGAGE 2022, LNCS 13862, pp. 67–79, 2023.
https://doi.org/10.1007/978-3-031-30923-6_6

In addition, unlike an ellipse, a parabola and a hyperbola also pass through one, respectively two, *points at infinity*, also called *ideal points* or *improper points.* Hence, such points can also be considered waypoints in the presented conic fitting problem. Through the use of improper points in the real projective plane $\mathbb{RP}^2$, we can easily extend the GAC embedding of points in the plane $\mathbb{R}^2$ (accordingly called *proper points*) to express the improper points in terms of GAC and incorporate the use of improper points into the derived conic fitting algorithm.

Let us note that rigorous proofs of statements are mostly omitted throughout the paper, as the focus is placed on the results.

2 Fitting in GAC – Without Given Waypoints

GAC constitutes a Clifford algebra $Cl(5,3)$ with embedding $C : \mathbb{R}^2 \to \mathbb{R}^{5,3}$ of a point $\mathbf{x} = xe_1 + ye_2$ of the plane $\mathbb{R}^2$ defined as

$$C(x,y) = \bar{n}_+ + xe_1 + ye_2 + \frac{1}{2}(x^2+y^2)n_+ + \frac{1}{2}(x^2-y^2)n_- + xyn_\times. \tag{1}$$

Consequently, the inner product null space (IPNS) representation of a general conic section Q in GAC is given by

$$Q_I = \bar{v}^\times \bar{n}_\times + \bar{v}^- \bar{n}_- + \bar{v}^+ \bar{n}_+ + v^1 e_1 + v^2 e_2 + v^+ n_+. \tag{2}$$

Also, we can represent a point embedded into GAC (using operator (1)) in vector form as

$$P_I = \begin{pmatrix} 0 & 0 & 1 & x & y & \frac{1}{2}(x^2+y^2) & \frac{1}{2}(x^2-y^2) & xy \end{pmatrix}^T \tag{3}$$

and a GAC conic section (2) as a vector

$$Q_I = \begin{pmatrix} \bar{v}^\times & \bar{v}^- & \bar{v}^+ & v^1 & v^2 & v^+ & 0 & 0 \end{pmatrix}^T. \tag{4}$$

Moreover, an associated bilinear form of the inner product of vectors in GAC is given by the matrix

$$B = \begin{pmatrix} 0_{3\times3} & 0_{3\times2} & -I_3 \\ 0_{2\times3} & E_2 & 0_{2\times3} \\ -I_3 & 0_{3\times2} & 0_{3\times3} \end{pmatrix}, \quad \text{where } I_3 = \begin{pmatrix} 0 & 0 & 1 \\ 0 & 1 & 0 \\ 1 & 0 & 0 \end{pmatrix} \text{ and } E_2 = \begin{pmatrix} 1 & 0 \\ 0 & 1 \end{pmatrix}. \tag{5}$$

Hrdina, Návrat and Vašík, [5], define a conic fitting problem in terms of GAC as follows: For a conic represented by a vector Q of the form (4) and for N_D given data points represented by vectors P_i of the form (3), we assume the objective function to be given by

$$Q \mapsto \sum_i (P_i \cdot Q)^2, \tag{6}$$

where $\cdot$ denotes the inner product between vectors in GAC. The conic best fitting the points with respect to this function is represented by the Q that minimises this function. To avoid the geometrically meaningless minimum $Q=0$, the authors of [5] consider the natural normalisation constraint

$$Q^2 = 1. \tag{7}$$

Using the matrix of the bilinear form (5), the objective function (6) then reads

$$Q \mapsto \sum_i (P_i B Q)^2 = \sum_i Q^T B P_i P_i^T B Q = Q^T P Q,$$

and thus, it is a quadratic form on $\mathbb{R}^{5,3}$ with the matrix

$$P = \sum_i B P_i P_i^T B.$$

To formulate the solution to the optimisation problem (6), (7), it is advantageous to decompose the matrix P into the following blocks:

$$P = \begin{pmatrix} P_0 & P_1 & 0 \\ P_1^T & P_c & 0 \\ 0 & 0 & 0 \end{pmatrix},$$

where P_0 is a 2×2 matrix, P_1 is a 2×4 matrix and P_c is a 4×4 matrix. The subscript c denotes that this block corresponds to the CRA part in GAC. Similarly, B_c denotes the middle 4×4 part of B, (5), and it coincides with the matrix of the inner product in CRA, [3]. Using the defined vectors and matrices, the desired solution is acquired according to the following proposition.

Proposition 1. *The solution to the optimisation problem* (6), (7) *for conic fitting in GAC is given by* $Q = \begin{pmatrix} w^T & v^T & 0 \end{pmatrix}^T$, *where* $v = \begin{pmatrix} \bar{v}^+ & v^1 & v^2 & v^+ \end{pmatrix}^T$ *is an eigenvector corresponding to the minimal non-negative eigenvalue of the operator*

$$P_{con} = B_c(P_c - P_1^T P_0^{-1} P_1)$$

and $w = \begin{pmatrix} \bar{v}^\times & \bar{v}^- \end{pmatrix}^T$ *is a vector acquired as*

$$w = -P_0^{-1} P_1 v.$$

The proof of Proposition 1 and the corresponding algorithm implemented in MATLAB together with the experimental results can be found in [5,7,8].

3 Proper and Improper Points in GAC

To shed light on the meaning of improper points, let us in short recall the concept of the *real projective plane* $\mathbb{RP}^2$, [11]:

Let $\mathbb{E} = (\mathcal{P}_\mathbb{E}, \mathcal{L}_\mathbb{E}, \mathcal{I}_\mathbb{E})$ be the usual Euclidean plane with points $\mathcal{P}_\mathbb{E} = \mathbb{R}^2$, lines $\mathcal{L}_\mathbb{E}$, and the usual incidence relation $\mathcal{I}_\mathbb{E} \subseteq \mathcal{P}_\mathbb{E} \times \mathcal{L}_\mathbb{E}$ of the Euclidean plane. We can easily extend the Euclidean plane to the real projective plane by including elements at infinity.

Now, for a line l, let us consider the equivalence class $[l]$ of all the lines that are parallel to l. For each such equivalence class, we define a new point $p_{[l]}$ that serves as an *improper point* at which all the parallels contained in the equivalence class $[l]$ intersect. Furthermore, we define one line at infinity l_∞ on which all the points $p_{[l]}$ lie. Consequently, the *real projective plane* can be defined as follows:

Definition 1. *Real projective plane* $\mathbb{RP}^2$ *is a triple* $(\mathcal{P}, \mathcal{L}, \mathcal{I})$*, where*

- $\mathcal{P} = \mathcal{P}_\mathbb{E} \cup \{p_{[l]} | l \in \mathcal{L}_\mathbb{E}\}$,
- $\mathcal{L} = \mathcal{L}_\mathbb{E} \cup l_\infty$,
- $\mathcal{I} = \mathcal{I}_\mathbb{E} \cup \{(p_{[l]}, l) | l \in \mathcal{L}_\mathbb{E}\} \cup \{(p_{[l]}, l_\infty) | l \in \mathcal{L}_\mathbb{E}\}$.

A sketch of three distinct bundles (equivalence classes) of parallel lines can be seen in Fig. 1—on the left, we can see the lines depicted in plane $\mathbb{R}^2$; on the right, we can see the situation in $\mathbb{RP}^2$: all the parallels from one equivalence class intersect in one common improper point on the line at infinity, which can be imagined as a circle with an infinite radius where the antipodal points are assumed to be identical (so every bundle of parallels really intersect in one point, not in two points, as it would seem at first glance).

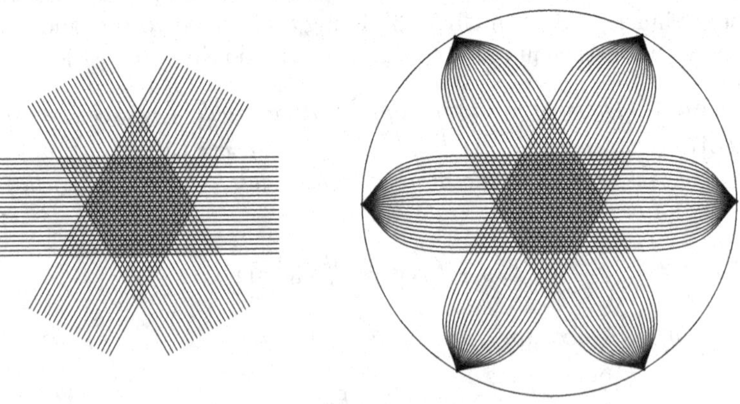

Fig. 1. Parallel lines and their common improper points in $\mathbb{RP}^2$ (taken from [11])

In contrast with the proper points, the improper points are not part of the plane $\mathbb{R}^2$, so an improper point is usually depicted using the direction (vector) of the parallel lines that meet at it, as illustrated in Fig. 2.

Having a notion of improper points, we can state that a parabola has one improper point corresponding to the direction of its axis of symmetry, while a hyperbola has two improper points in the directions of its asymptotes (see Fig. 2).

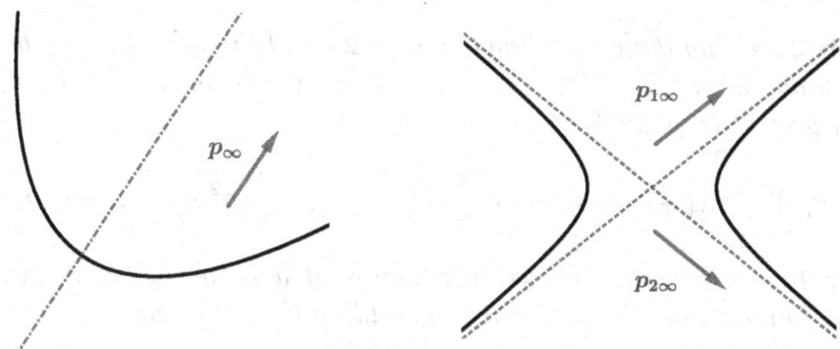

Fig. 2. Improper points of parabola and hyperbola

3.1 Homogeneous Coordinates

The concept of $\mathbb{RP}^2$ allows us to distinguish between proper and improper points and, at the same time, to carry out computations with them as if they were not different at all. Using *homogeneous coordinates*, every point of $\mathbb{RP}^2$ (proper or improper) can be represented as a line in $\mathbb{R}^3$ according to the following definition.

Definition 2. *Let* $\mathbf{x} = (x, y), \mathbf{x} \in \mathbb{RP}^2$, *be a proper point, then its homogeneous coordinates are*

$$\mathbf{x} = k(x, y, 1), \qquad k \in \mathbb{R} \setminus \{0\},$$

while the homogeneous coordinates of an improper point $\mathbf{x}_\infty = (s, t), \mathbf{x}_\infty \in \mathbb{RP}^2$, *are*

$$\mathbf{x}_\infty = k(s, t, 0), \qquad k \in \mathbb{R} \setminus \{0\}.$$

Note that the triple $(0, 0, 0)$ does not represent any point of $\mathbb{RP}^2$.

Remark 1. While the homogeneous coordinates (a, b, c) represent the same point in $\mathbb{RP}^2$ as a triple $k(a, b, c)$ for every non-zero k, it is very advantageous to take $k = 1$ when assigning homogeneous coordinates to the points of the real projective plane. For the sake of computational simplicity, we will assume that the homogeneous coordinates of a proper point $\mathbf{x} = (x, y)$ and of an improper point $\mathbf{x}_\infty = (s, t)$, respectively, are given by the mapping

$$(x, y) \mapsto (x, y, 1),$$
$$(s, t) \mapsto (s, t, 0).$$

3.2 Projectivisation of GAC

With homogeneous coordinates in hand we will show it is possible to use GAC for the representation of proper and improper points alike. Such a representation can be achieved by extending the domain of the point embedding $C : \mathbb{R}^2 \to \mathbb{R}^{5,3}$ (see (1)) to $\mathbb{RP}^2$ in the following way.

Definition 3. *Using the embedding* $C : \mathbb{R}^2 \to \mathbb{R}^{5,3}$ *of the form* (1), *we define the projective embedding* $C\mathbb{P} : \mathbb{RP}^2 \to \mathbb{R}^{5,3}$ *of a point* $\mathbf{p} = (a, b, c)$, $(a, b, c) \neq (0, 0, 0)$, *in the real projective plane* $\mathbb{RP}^2$ *as*

$$C\mathbb{P}(a,b,c) = c^2\bar{n}_+ + ace_1 + bce_2 + \frac{1}{2}(a^2+b^2)n_+ + \frac{1}{2}(a^2-b^2)n_- + abn_\times. \quad (8)$$

Corollary 1. *Since homogeneous coordinates of a proper point* $\mathbf{x} = (x, y)$ *are* $(x, y, 1)$, *it follows that the projective embedding* $C\mathbb{P}$, (8), *maps a proper point into GAC in the same way as the embedding* C*:*

$$C\mathbb{P}(x,y,1) \equiv C(x,y) = \bar{n}_+ + xe_1 + ye_2 + \frac{1}{2}(x^2+y^2)n_+ + \frac{1}{2}(x^2-y^2)n_- + xyn_\times.$$

On the other hand, an improper point $\mathbf{x}_\infty = (s, t)$ *with homogeneous coordinates* $(s, t, 0)$ *is mapped by* $C\mathbb{P}$ *in a simpler way:*

$$C\mathbb{P}(s,t,0) = \frac{1}{2}(s^2+t^2)n_+ + \frac{1}{2}(s^2-t^2)n_- + stn_\times. \quad (9)$$

Consequently, in addition to the IPNS vector representation of a proper point $\mathbf{x} = (x, y)$ *embedded into GAC of the form* (3), *we can also define such a vector representation of an improper point* $\mathbf{x}_\infty = (s, t)$ *according to* (9) *as*

$$P_{\infty I} = \begin{pmatrix} 0 & 0 & 0 & 0 & 0 & \tfrac{1}{2}(s^2+t^2) & \tfrac{1}{2}(s^2-t^2) & st \end{pmatrix}^T. \quad (10)$$

4 Fitting in GAC - With Given Waypoints

To successfully fit a conic among the data points as tightly as possible while ensuring that the fitted conic will pass through the prescribed waypoints, we must exploit the structure of GAC and the vector-matrix description of the conic fitting problem as presented in Sect. 2. Fortunately, as will be shown, a solution to the conic fitting problem with waypoints can be formulated in a way similar to the solution to conic fitting without waypoints.

As in the case of conic fitting without waypoints, we are given N_D data points represented by vectors P_i of the form (3) and we seek a conic represented by a vector Q of the form (4) minimising the objective function (6) while fulfilling the normalisation constraint (7). Let us note that we assume all the data points P_i to be *proper* points, since fitting among the improper points without actually passing through them would be geometrically meaningless.

In addition, we demand the fitted conic to pass through N_W waypoints W_j that can be either proper or improper, unlike the data points P_i, and therefore each of them must be either of the form (3) or (10). For a conic Q to pass through the waypoint W, their inner product must be zero, i.e.

$$W \cdot Q = 0.$$

Using the matrix Ψ of waypoints W_j, where the j-th column is waypoint W_j, we can formulate the condition of conic Q passing through all the waypoints W_j as

$$\Psi \cdot Q = 0. \tag{11}$$

To successfully reach the solution to the given optimisation problem, we define a matrix B_0 and the decomposition of matrix Ψ of waypoints as:

$$B_0 = \begin{pmatrix} 0 & -1 \\ -1 & 0 \end{pmatrix}, \quad \Psi = \left(\begin{array}{c|c|c|c} W_1 & W_2 & \cdots & W_{N_W} \end{array} \right) = \begin{pmatrix} 0 \\ \hline \Xi \\ \hline X \end{pmatrix},$$

where 0 stands for the zero matrix of type $2 \times N_W$, Ξ is of type $4 \times N_W$ and X has a size of $2 \times N_W$. Moreover, we define the matrix

$$B_w = \left(B_0^T X\right)^{+T} \Xi^T B_c, \tag{12}$$

where "+" stands for the Moore-Penrose pseudoinverse, since the matrix $B_0^T X$ is generally not square.

Consequently, we can formulate a statement similar to Proposition 1 and, again, reach the solution using an eigenproblem.

Proposition 2. *The solution to the optimisation problem* (6), (7), (11) *for conic fitting in GAC is given by* $Q = \left(w^T \ \ v^T \ \ 0\right)^T$, *where* $v = \left(\bar{v}^+ \ \ v^1 \ \ v^2 \ \ v^+\right)^T$ *is an eigenvector corresponding to the minimal non-negative eigenvalue of the operator*

$$P_{con}^W = B_c \left[B_w^T P_0 B_w - \left(B_w^T P_1 + P_1^T B_w\right) + P_c\right]$$

and $w = \left(\bar{v}^\times \ \ \bar{v}^-\right)^T$ *is a vector acquired as*

$$w = -B_w v.$$

4.1 Implementation

Below, we summarise an algorithm for conic fitting with given waypoints implemented as a MATLAB function. Let us note that the reduced forms of some vectors and matrices were employed to avoid a few unnecessary computations with zero elements, similarly to the conic fitting algorithms in [7,8].

Algorithm QW

Inputs:

a, b, c column vectors of x, y, z homogeneous coordinates of waypoints (if a point is proper, take $z = 1$, if improper, $z = 0$; see Remark 1)
px, py column vectors of x, y coordinates of data points

Outputs:

Conic fitted conic in form (4)
obj_function value of objective function (6) for fitted conic

---

```
function [Conic, obj_function] = QW(a,b,c,px,py)
ND = length(px);
s = sign(c);

Xi = [s a.*s b.*s 1/2*(a.^2+b.^2)]';
Chi = [1/2*(a.^2-b.^2) a.*b]';

B = zeros(6);
I3 = [0 0 1;0 1 0; 1 0 0];
B(1:3,4:6) = -I3;
B(4:5,2:3) = eye(2);
B(6,1) = -1;
Bc = B(3:6,1:4);
B0 = [0 -1;
    -1 0];
Bw = (pinv(B0'*Chi))'*Xi'*Bc;

D = ones(6,ND);
D(2,:) = px;
D(3,:) = py;
D(4,:) = 1/2*(px.^2+py.^2);
D(5,:) = 1/2*(px.^2-py.^2);
D(6,:) = px.*py;

P = 1/ND*B*(D*D')*B';
Pc = P(3:6,3:6);
P0 = P(1:2,1:2);
P1 = P(1:2,3:6);

PWcon = Bc*(Bw'*P0*Bw-(Bw'*P1+P1'*Bw)+Pc);
[EV,ED] = eig(PWcon);
EW = diag(ED);

k_opt = find(EW == min(EW(EW>0)));
v_opt = EV(:,k_opt);

kappa = v_opt'*Bc*v_opt;
v_opt = 1/sqrt(kappa)*v_opt;

w = -Bw*v_opt;

Conic = [w;v_opt;0;0];
obj_function = Conic(1:6)'*P*Conic(1:6);
end
```

4.2 Number of Waypoints and Degrees of Freedom of the Conic

It is widely known that a general conic is uniquely determined by five points of $\mathbb{RP}^2$ or, more precisely, it is uniquely determined by five points when no four of them lie on the same line. Moreover, when no three points out of these five lie on the same line, the conic passing through them is not only uniquely determined, but also regular [6, 10]. This well corresponds with the fact that a conic has five degrees of freedom.

Taking such knowledge into account, it is hence not possible to fit a conic through more than a few waypoints while minimising the objective function (6) and satisfy the normalisation constraint (7).

As already suggested, conic fitting through five or more waypoints would make no sense in our problem. Therefore, a maximum of four waypoints comes into play. Since we can use either proper or improper waypoints or even a combination of both types of waypoints, we get 14 cases of what waypoints to employ in total, as can be seen in Table 1.

Table 1. Combinations of waypoints by type and total number

	1A	1B	2A	2B	2C	3A	3B	3C	3D	4A	4B	4C	4D	4E
proper	1	0	2	1	0	3	2	1	0	4	3	2	1	0
improper	0	1	0	1	2	0	1	2	3	0	1	2	3	4

It can be further shown that the presented algorithm works *safely* for two waypoints at most; the reasons for this are various. Since the normalisation constraint (7) makes some of conic elements dependent, at least one degree of freedom is lost. Moreover, some combinations are geometrically unreasonable, e.g. the case 3D, where the conic is required to pass through 3 improper waypoints. Finally, since the waypoints constraint (11) may imply an overdetermined system of equations, our solution cannot generally satisfy all the equations (not even after using the pseudoinverse in (12)).

5 Experimental Results

We applied the conic fitting algorithm with given waypoint(s) on three different datasets listed in Table 2 (the location of the points in each dataset was deliberately chosen to resemble the shape of a regular conic; hence, the corresponding datasets were named *elliptical*, *parabolic* and *hyperbolic*, respectively).

Moreover, by varying the total number and types of the waypoints used, we offer 8 selected cases of fits from Table 1. The particular waypoints used in the experiments are listed in Table 3.

Table 2. Datasets used

x_i	3	4	3	0	−1	−3	−4	−3	−1	−2	−2	−1	−1	0	1	3	5	7	−6	−4	−4	−3	−3	1	2	2	4
y_i	−1	1	2	3	3	2	−1	−3	−4	0	3	−2	6	8	−2	−2	−1	0	1	2	−3	1	−1	5	4	7	4
	elliptical									parabolic									hyperbolic								

In Fig. 3 we can see four cases of the elliptical dataset fitted by an ellipse, each fit as a result of fitting a conic with one more proper waypoint than in the preceding subfigure (see Table 3). The outcome of this experiment corresponds with the commentary on degrees of freedom given in Subsect. 4.2: in cases 1A and 2A the conic passes through all the waypoints, while in cases 3A and 4A it does not because the associated systems of equations are overdetermined.

Table 3. Waypoints used and the corresponding cases

x_j	2	3	−4	−2	4	6	1	−7	1	5	1	6
y_j	−3	3	1	−4	5	1	2	1	2	4	2	1
z_j	1	1	1	1	0	0	0	1	0	1	0	0
	1A-4A				1B	2C		2B		3C		

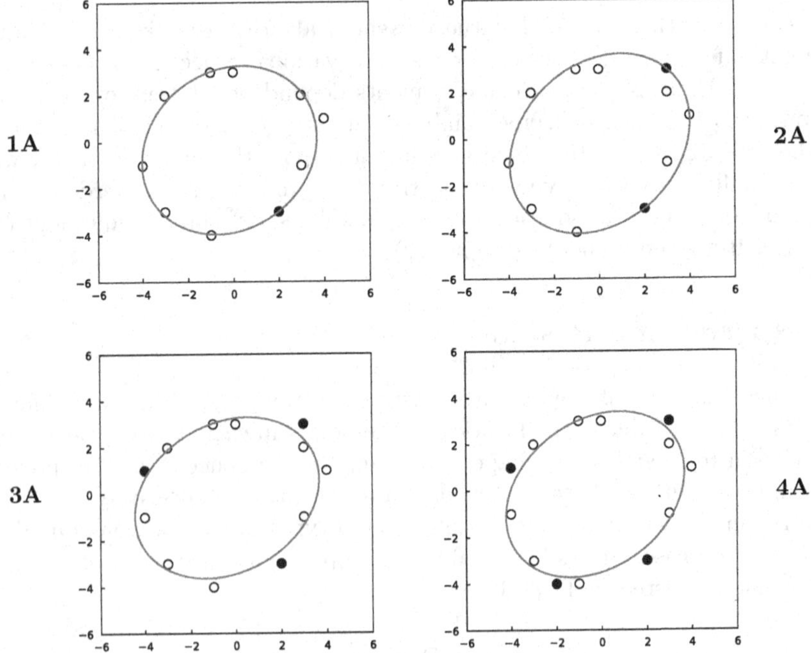

Fig. 3. Conic fits using 1–4 proper waypoints

The second experiment summarised in Fig. 4 was intended to show the behaviour of the algorithm in the cases when only improper waypoints are prescribed (here, we limited ourselves to geometrically meaningful cases, i.e. two improper waypoints at most).

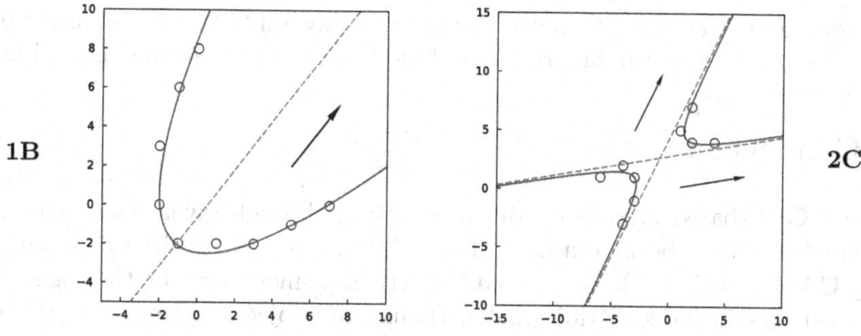

Fig. 4. Conic fits passing through 1 and 2 improper points respectively

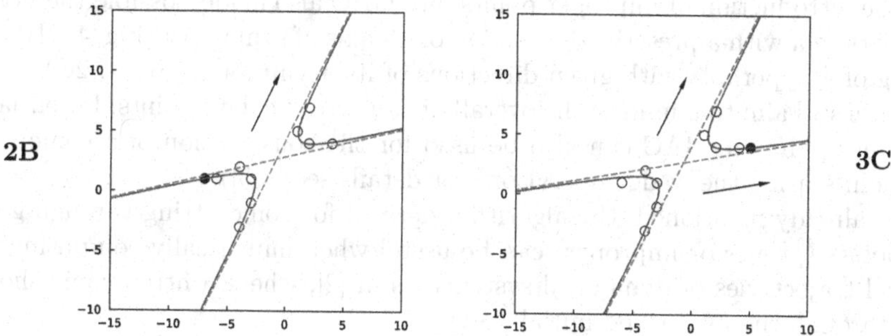

Fig. 5. Conic fits with both proper and improper waypoints

In the case 1B, where one improper waypoint was given, a parabola with the axis of symmetry passing through the same improper waypoint was fitted. While this result may not be surprising, it is actually far from obvious. Even though the dataset was meant to resemble a parabolic shape and every parabola has one improper point, the algorithm could still have fitted a hyperbola with one asymptote passing through a given waypoint and the other one might have been a result of the optimisation process. In fact, it can be shown that the employed fitting algorithm with only one improper waypoint and no proper waypoint fits a conic with *exactly one* improper point, i.e. a parabola.

Similarly, in the case 2C, a hyperbola passing through *exactly two* given improper waypoints is fitted.

Finally, let us explore the situations when both proper and improper waypoints are fitted at the same time (see Fig. 5). While the case 2B might have turned out the same as the case 1B, where one improper waypoint was used as well, the additional proper waypoint helped to create a hyperbolic fit passing through both given waypoints (moreover, it can be shown that the fit through one proper and one improper point results in a parabola using very specific configurations of points only). Let us also note that the second improper point lying in the direction of the second asymptote was a mere result of optimisation.

Even though the case 3C makes use of three waypoints and hence, as already indicated, is not very reliable, there are functioning cases, like the one in Fig. 5.

6 Conclusion

A novel GAC-based algorithm for conic fitting through given waypoints was presented and thus became a part of a wider group of conic fitting algorithms using GAC (see [5,7,8]). Also, a MATLAB implementation of the algorithm described was included. Additionally, thanks to projectivisation of GAC, the expression of improper points in terms of GAC was found and, consequently, the improper waypoints could be used in the algorithm as well.

The introduction of improper points into GAC also made possible the fitting of a parabola with a prescribed direction of its axis of symmetry (Fig. 4, 1B) and fitting of a hyperbola with given directions of its asymptotes (Fig. 4, 2C), while both fits still aim to minimise the overall distance to the data points. Potentially, improper points in GAC can also be used for the construction of a conic from five points using the wedge operation (for details see [4,5]).

As already mentioned, the algorithm derived for conic fitting through given waypoints (proper or improper) can be useful when numerically computing the conical trajectories of dynamical systems, as in [2], where a fitted conic should pass through the prescribed initial point.

Finally, it was experimentally shown that the algorithm works safely when two waypoints at most are employed, while more waypoints may cause the fitted conic to miss the waypoints. Precise reasoning for this behaviour, together with a more comprehensive analysis of fitting through waypoints, will be the subject of further research.

References

1. Byrtus, R., Derevianko, A., Vašík, P., Hildenbrand, D., Steinmetz, C.: On specific conic intersections in GAC and symbolic calculations in GAALOPWeb. Adv. Appl. Clifford Algebras **32**(1) (2021). https://doi.org/10.1007/s00006-021-01182-z. ISSN 0188-7009
2. Derevianko, A.I., Vašík, P.: Solver-free optimal control for linear dynamical switched system by means of geometric algebra. Math. Meth. Appl. Sci., 1–15 (2022). https://doi.org/10.1002/mma.8752
3. Hildenbrand, D.: Introduction to Geometric Algebra Computing. CRC Press, Taylor & Francis Group (2019)
4. Hrdina, J., Návrat, A., Vašík, P.: Geometric algebra for conics. Adv. Appl. Clifford Algebras **28**, 66 (2018). https://doi.org/10.1007/s00006-018-0879-2
5. Hrdina, J., Návrat, A., Vašík, P.: Conic fitting in geometric algebra setting. Adv. Appl. Clifford Algebras **29**, 72 (2019). https://doi.org/10.1007/s00006-019-0989-5
6. Korn, G.A., Korn, T.M.: Mathematical Handbook for Scientists and Engineers. McGraw-Hill Book Company (1961)

7. Loučka, P., Vašík, P.: Algorithms for multi-conditioned conic fitting in geometric algebra for conics. In: Magnenat-Thalmann, N., et al. (eds.) CGI 2021. LNCS, vol. 13002, pp. 645–657. Springer, Cham (2021). https://doi.org/10.1007/978-3-030-89029-2_48
8. Loučka, P., Vašík, P.: On multi-conditioned conic fitting in Geometric algebra for conics. arXiv:2103.14072 [math.NA]
9. Perwass, Ch.: Geometric Algebra with Applications in Engineering. Springer, Heidelberg (2009)
10. Pamfilos, P.: A gallery of conics by five elements. Forum Geometricorum **14**, 295–348 (2014)
11. Richter-Gebert, J.: Perspectives On Projective Geometry: A Guided Tour Through Real and Complex Geometry. Springer, Heidelberg (2016)

Modelling Proteins and Cities

Using a Graph Transformer Network to Predict 3D Coordinates of Proteins via Geometric Algebra Modelling

Alberto Pepe[1(✉)], Joan Lasenby[1], and Pablo Chacón[2]

[1] Signal Processing and Communications Group, Cambridge University Engineering Department, Trumpington Street, Cambridge CB2 1PZ, UK
{ap2219,jl221}@cam.ac.uk

[2] Chacon Lab, Rocasolano Physical Chemistry Institute, 28006 Madrid, Spain

Abstract. The state of the art in protein structure prediction (PSP) is currently achieved by complex deep learning pipelines that require several input features. In this paper, we demonstrate the relevance of Geometric Algebra (GA) for modelling protein features in PSP. We do so by proposing a novel GA metric based on the relative orientations of amino acid residues. We then employ this metric as an additional input feature to a Graph Transformer (GT) to aid the prediction of the 3D coordinates of a protein. Adding this GA-based orientational information improves the accuracy of the predicted coordinates even after few learning iterations and on a small dataset.

Keywords: protein structure prediction · 3D modelling · geometric algebra · graph transformer

1 Introduction

The last Critical Assessment of Protein Structure Prediction (CASP14) was won by AlphaFold 2, reaching an unprecedented global distance test (GDT) score of above 90% in almost 70% of the proteins in the CASP dataset [1–3]. AlphaFold 2 confirmed that deep learning (DL) is the most successful approach for PSP, and significantly cheaper and faster than experimental techniques [4–6].

A typical DL-based PSP pipeline is generally composed of several cascaded neural networks, whose end goal is the prediction of 3D coordinates of some of the atoms in the protein backbone [7]. In recent literature, Transformer networks have been proven to be particularly suitable for this task [1,7,12]. Transformer networks are sequence-to-sequence models first introduced in [8], and have found widespread application in fields including speech synthesis [9], semantic correspondence [10] and trajectory forecasting [11].

E. Hitzer et al. (Eds.): ENGAGE 2022, LNCS 13862, pp. 83–95, 2023.
https://doi.org/10.1007/978-3-031-30923-6_7

In PSP, for example, two of the seven networks employed in [7] are Transformer networks to predict and refine the coordinates of the backbone atoms, respectively. Similarly, a multiple sequence alignment (MSA) Transformer followed by a GT has been employed to predict 3D coordinates starting from the protein's sequence of amino acids in [12].

The 3D coordinates are predicted by training the network on several biological and chemical features of the protein. These features are extracted starting from its amino acid sequence (or primary structure) [14]. It has been shown that the interresidue distances (e.g. distance between amino acid pairs), the secondary structures of the proteins (e.g. the folding patterns such as helices, sheets or turns), as well as some measure of the orientation between amino acids (e.g. angle maps) are among the most relevant features when learning accurate 3D coordinates [7,13,14].

GA is a suitable candidate to represent the features mentioned above due to its intuitive handling of geometrical objects and operations on them [15,16]. GA has already found some applications in protein modelling, especially in the molecular distance problem [17,18], but to the best of our knowledge there has not been an effort to employ GA modelling for PSP.

The goal of this paper is hence to (1) employ GA to model a protein and capture information about the orientation of the amino acids and (2) use this information as a feature in a GT network. The motivations of using GA are that: (1) GA easily deals with geometrical objects such as planes, which naturally occur in the protein geometry (2) our GA feature is more compact compared with torsion and valence angles, which also grasp orientational information, but are more than one and asymmetrical, as seen in [13] and (3) it has a clear physical meaning, since it is related to secondary structures (see Sect. 2.1).

The rest of the paper is structured as follows: in Sect. 2 protein modelling through GA and graphs are presented, in Sect. 3 the learning architecture is introduced, in Sect. 4 results are presented and in Sect. 5 conclusions are drawn.

2 Modelling Proteins

2.1 Proteins as Rigid Bodies

The atoms in the protein backbone determine its overall shape. Each amino acid is bonded to an α-carbon (C_α), which is preceded by a nitrogen (N) atom and followed by a carbon (C) atom. Hence, there is a one-to-one correspondence between an amino acid i and a triplet $\{N, C_\alpha, C\}_i$.

Each $\{N, C_\alpha, C\}$ triplet lies on a plane. We can take advantage of this information and associate each triplet i with a plane Π_i in Conformal Geometric Algebra (CGA): let A_i, B_i and C_i be the CGA representations of the Euclidean coordinates of the atoms $\{N, C_\alpha, C\}_i$. Π_i can be then computed as the 4-blade:

$$\Pi_i = A_i \wedge B_i \wedge C_i \wedge n_\infty \tag{1}$$

where $n_\infty = e + \bar{e}$, with $e^2 = +1, \bar{e} = -1$ being two basis vector of $\mathcal{G}_{4,1,0}$ and $\wedge$ denoting the outer product.

In this way, a protein is modelled as a collection of planes not too dissimar to the gas of 3D rigid bodies of AlphaFold 2 [19] (see Fig. 1).

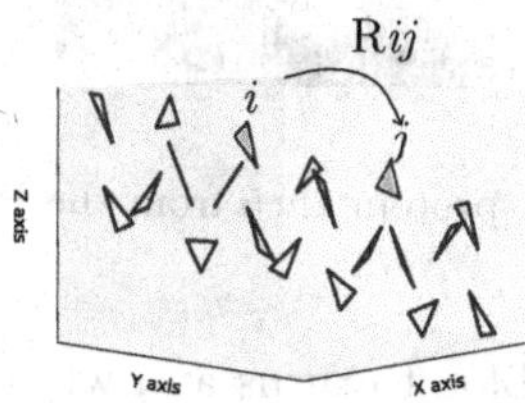

Fig. 1. A toy helix protein as a collection of planes. $\Pi_j = R_{ij}\Pi_i\tilde{R}_{ij}$

For each pair of planes Π_i, Π_j we can then form a rotor that rotates Π_i into Π_j as presented in [20]:

$$R_{ij} = \frac{1}{\sqrt{\langle\xi\rangle_0}}(1 - \Pi_i\Pi_j) \tag{2}$$

where $\xi = 2 - (\Pi_i\Pi_j + \Pi_j\Pi_i)$ and $\langle\cdot\rangle$ is the grade projector operator.

We now use the cost function $\mathscr{C}_\lambda(R)$ as defined in [21] that quantifies the variation of R from the identity. $\mathscr{C}_\lambda(R)$ is a weighted sum of a translational and a rotational term:

$$\mathscr{C}_{\lambda_1\lambda_2}(R) = \lambda_1\langle R_\parallel\tilde{R}_\parallel\rangle_0 + \lambda_2\langle(R_\perp - 1)(\tilde{R}_\perp - 1)\rangle_0 \tag{3}$$

in which the translational error is represented by $R_\parallel = R \cdot e$, and the rotational error by $\langle(R_\perp - 1)(\tilde{R}_\perp - 1)\rangle_0 = \langle(R - 1)(\tilde{R} - 1)\rangle_0$. As we are interested in an orientational feature, we will focus exclusively on the rotational part (i.e. $\lambda_1 = 0, \lambda_2 = 1$).

Since each amino acid can be associated with a plane, and each pair of planes can be associated with a rotor and eventually to a cost, we can then build an $N \times N$ matrix $\mathbf{M}$ as follows:

$$\mathbf{M}_{ij} = \begin{cases} C_{\lambda_1\lambda_2}(R_{ij}) & \text{if } d_{ij} < 15\,\text{Å} \\ 0 & \text{otherwise} \end{cases} \tag{4}$$

where N is the amino acid sequence length and d_{ij} is the Euclidean distance between the C_α of residues i, j measured in Å. We call $\mathbf{M}$ a "cost map". An example of a cost map is given in Fig. 2.

It is possible to establish a relationship between the secondary structure and the patterns in the cost maps. By secondary structure we refer to local folding patterns of a protein, including α-helices, β-sheets or turns. We illustrate this relationship by assigning an arbitrary colour to each secondary structure: red to

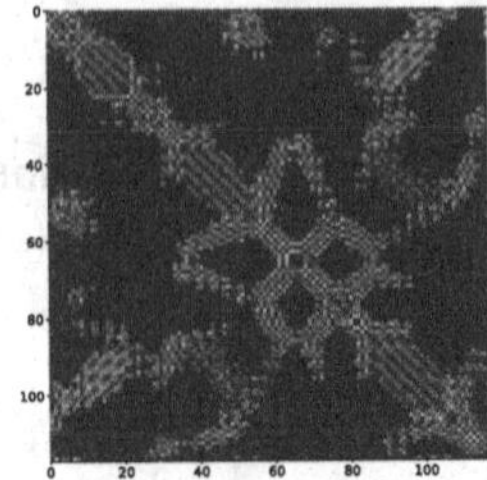

Fig. 2. Cost map for protein 2hc5 from the PDB database [24]

α-helices, green to β-sheets, blue to turns and white to all the others. In Fig. 3 we see how the same colour patches have almost identical cost map patterns.

To the best of our knowledge, this is the first example of a single orientational map based on GA that matches the secondary structures.

Fig. 3. Colour coded secondary structures overlapping the cost map of protein 2hc5. (Color figure online)

2.2 Proteins as Graphs

It is also possible to represent a protein as a heterogeneous graph $\mathcal{G}(V, E)$ with V and E being its set of nodes and edges, respectively. By heterogeneous graph we refer to a graph with different types of nodes and edges. If $|V| = N$ is the total number of nodes, the graph can be described as a set of adjacency matrices for each of the K edge types, i.e. $\{A_k\}_{k=1}^{K}$, where $A_k \in \mathbb{R}^{N \times N}$, or in tensor form $\mathbf{A} \in \mathbb{R}^{N \times N \times K}$. Along with $\mathbf{A}$, we can also define a feature matrix $X \in \mathbb{R}^{N \times D}$, where D is the dimensionality of the features, or equivalently we can say there are D node types.

For our experiment, we employed the PDNET dataset and recast it in graph form [14]. PDNET is composed of a stack of 57 $N \times N$ channels for each of its proteins. We can hence associate each pairwise feature with an edge type and each per-amino acid feature with a node type. Of the 57 channels, 3 of them correspond to 3 pairwise features (FreeCon, CCMPred and potential). To these 3 we added distance maps (defined as $\mathbf{D}_{ij} = d_{ij}$, where $d_{ij} = \|T_i - T_j\|_2$, with

$T \in \mathbb{R}^{N\times 3}$ being the ground truth coordinates of the C_α atoms of the protein) and cost maps $\mathbf{M}$, to obtain a total of $K = 5$ pairwise maps of size $N \times N$, with N being the protein chain length. They correspond to the edges of the protein graph, i.e. the adjacency matrices $\mathbf{A} \in \mathbb{R}^{N\times N\times K=5}$. The remaining 54 channels are the matrices and their transposes of the remaining 4 features (SA, PSSM, SS, entropy), which are associated with a single amino acid. Ignoring the transposed matrices, we are left with 27 channels, which can be manipulated and arranged in a feature matrix $X \in \mathbb{R}^{N\times D=27}$.

The input to the architecture is then given by the pair of tensors $\{\mathbf{A}, X\}^{(i)}$ for each protein i in the dataset.

3 Architecture

The end-to-end architecture, derived from [12], is composed of two parts: (1) a GT and (2) 3D projector. A summary of the architecture is shown in Fig. 4. We omitted the MSA Transformer of [12] as the employed dataset allows us to directly perform node and edge embedding on its features.

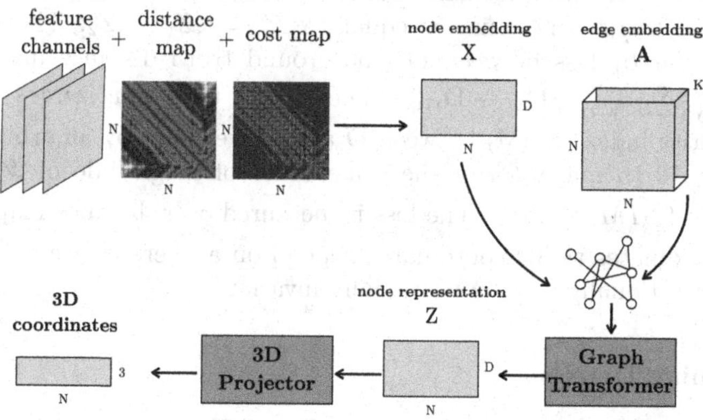

Fig. 4. The employed architecture

3.1 Graph Transformer

The GT has been implemented as described in [22]. The goal of a GT is to learn informative meta-path within the graph, i.e. an ordered sequence of node types and edge types. The output of the l-th layer of a GT with C attention heads is a node representation with same dimensionality as X, i.e. $Z \in \mathbb{R}^{N\times D}$ which can be written as

$$Z^{(l)} = \bigoplus_{i=1}^{C} \sigma(\tilde{\Delta}_i^{-1} \tilde{A}_i^{(l)} XW) \tag{5}$$

where $\bigoplus$ is the concatenation operator, $\sigma(\cdot)$ is the sigmoid function, $\tilde{\Delta}_i$ is the degree matrix of $\tilde{A}_i^{(l)}$ (defined as $\Delta_{mm} = \sum_n A_{mn}$), X is the feature matrix, $W \in \mathbb{R}^{D\times D}$ is a trainable weight matrix and $\tilde{A}_i^{(l)} = A_i^{(l)} + I$, in which $A_i^{(l)}$ is the adjacency matrix from the i-th channel of the metapath tensor $\mathbf{A}^{(l)} \in \mathbb{R}^{N\times N\times C}$. $\mathbf{A}^{(l)}$ is evaluated as $\mathbf{A}^{(l)} = \Delta^{-1}\mathbf{Q}_1\mathbf{Q}_2$. $\mathbf{Q}_1$ and $\mathbf{Q}_2$, both $\in \mathbb{R}^{N\times N\times C}$, are two adjacency tensors selected according to $\mathbf{Q} = \varphi[\mathbf{A}; \zeta(\mathbf{W}_\varphi)]$, where $\mathbf{A} \in \mathbb{R}^{N\times N\times K}$ is the adjacency tensor, $\varphi(\cdot)$ is the convolution operator, $\zeta(\cdot)$ is the softmax function and $\mathbf{W}_\varphi \in \mathbb{R}^{C\times C\times K}$ are the weights of φ. Z contains the node representations from C different meta-path graphs.

3.2 3D Projector

The 3D projector is a simple fully connected layer obeying $P = Z^{(L)}W_P$, where $Z^{(L)}$ is the output of the L-th layer of the GT, $W_P \in \mathbb{R}^{D\times 3}$ is the weight matrix of the projector and $P \in \mathbb{R}^{N\times 3}$ are the 3D coordinates of the N C_α atoms in the protein chain.

To train the model, a distance map is evaluated for each protein from the predicted coordinates P as $\tilde{\mathbf{D}}_{ij} = d_{ij}$, where $d_{ij} = \|P_i - P_j\|_2$ is the Euclidean distance between the 3D coordinates of the i-th and j-th amino acid in P.

The total loss to minimize is equal to $\mathscr{L} = \mathscr{L}_1 + \mathscr{L}_2$. The first term *minimizes* the L_1 loss between $\mathbf{D}$ (the ground truth distance map) and $\tilde{\mathbf{D}}$, as $\mathscr{L}_1 = \frac{1}{N^2}\sum_i^N\sum_j^N \|\tilde{\mathbf{D}}_{ij} - \mathbf{D}_{ij}\|_1$. The second term *maximizes* the structural similarity index (SSIM) between $\mathbf{D}$ and $\tilde{\mathbf{D}}$ weighted by an arbitrary coefficient $\alpha = 10$ to make $\mathscr{L}_2$ of the same order of magnitude of $\mathscr{L}_1$, namely $\mathscr{L}_2 = \alpha\left(1 - SSIM\{\mathbf{D}, \tilde{\mathbf{D}}\}\right)$. The loss is measured over distance maps and not over 3D coordinates as 3D coordinates depend on a reference frame, while distances are rotationally and translationally invariant.

3.3 Training Details

We trained the model consisting of the GT and 3D projector on the PDNET dataset. The model consists of 108813 trainable parameters, of which 108648 of the GT and 165 of the projector. The train and validation sets are subsets of PDNET composed of 200 proteins each, while the test set contains 150 proteins. The optimizer has been set to Adam with exponentially decaying learning rate, with initial learning rate $\eta_0 = 1 \times 10^{-2}$ and decay rate per epoch $\gamma = 0.9$. The GT has $C = 4$ attention heads and $L = 3$ layers. The batch size has been fixed to $B = 1$ and the network has been trained for $E = 5$ epochs, for a total of 1000 training iterations.

Combinations of $\eta \in \{1\times10^{-1}, 1\times10^{-2}, 1\times10^{-3}, 3\times10^{-4}\}$, $E \in \{3,5,10\}$, $B \in \{1,50,100\}$, $L \in \{3,6,10\}$, $C \in \{1,4,5\}$ have also been implemented and tested, but the hyperparameters above were found to be optimal for our problem.

The code has been written as a Jupyter Notebook on Google Colaboratory, run on an NVIDIA Tesla K80 GPU and it uses PyTorch for the DL architecture, the Clifford library for GA operations [23] and the PDB Module of Biopython for handling protein data. The GT was derived from [12]. Scripts and datasets are available upon request to the authors.

4 Results

We trained the architecture and collected results for two cases: (1) *with* cost maps ($D = 27, K = 5$) and (2) *without* cost maps ($D = 27, K = 4$), to verify whether adding a single additional GA-based adjacency matrix A_k in our graph could provide an improvement.

From the predicted coordinates $P \in \mathbb{R}^{N\times3}$ we constructed distance maps $\tilde{\mathbf{D}} \in \mathbb{R}^{N\times N}$, and we then measured the mean absolute error (MAE) and SSIM between $\mathbf{D}$ and $\tilde{\mathbf{D}}$. The MAE and SSIM distributions are presented in Table 1, while the distributions and percentiles over the test set are visualized in Figs. 5 and 6 for the MAE and the SSIM, respectively.

Table 1. Metric between original and predicted distance maps. Results without costs are in parenthesis.

Set	Metric	Max	Mean	Min	Std
Train	SSIM	0.98 (0.90)	0.88 (0.43)	0.12 (−0.10)	0.10 (0.22)
Test	SSIM	0.99 (0.86)	0.88 (0.43)	0.38 (−0.14)	0.10 (0.25)
Train	MAE (Å)	27.9 (23.3)	6.09 (7.38)	2.16 (3.32)	2.69 (3.05)
Test	MAE (Å)	10.3 (12.8)	5.99 (7.02)	2.49 (3.58)	2.35 (1.91)

Note in Table 1 how the average SSIM doubles from 0.43 when coordinates are predicted without cost maps to 0.88 when coordinates are predicted with cost maps. Similarly, the average MAE decreases by 1.29 Å and 1.03 Å on the train and test sets, respectively, when we include cost maps. From Fig. 5 it can be seen that the median MAE of the test set is found to be at about 5 Å with costs maps and at about 7 Å without cost. The improvement introduced with cost maps is even more evident in Fig. 6, in which the median SSIM of the test set is >0.4 without cost maps and >0.8 with cost maps.

We then aligned P and T via singular value decomposition (SVD) (see Appendix A) and performed the GDT, and evaluated the GDT_TS (total score) and GDT_HA (half size) between predicted coordinates P and ground truth coordinates T, obtained from the Protein Data Bank (PDB) [24].

$$\text{GDT\_TS} = \frac{p_{<1\text{Å}} + p_{<2\text{Å}} + p_{<4\text{Å}} + p_{<8\text{Å}}}{4} \tag{6}$$

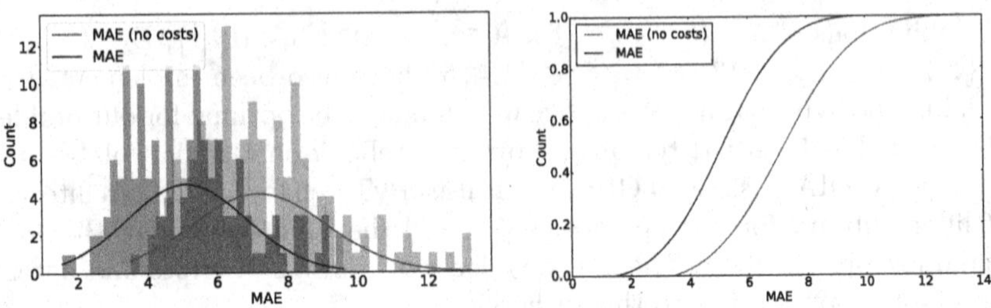

Fig. 5. MAE measured over the testing set. Distribution (left) and cumulative probability (right).

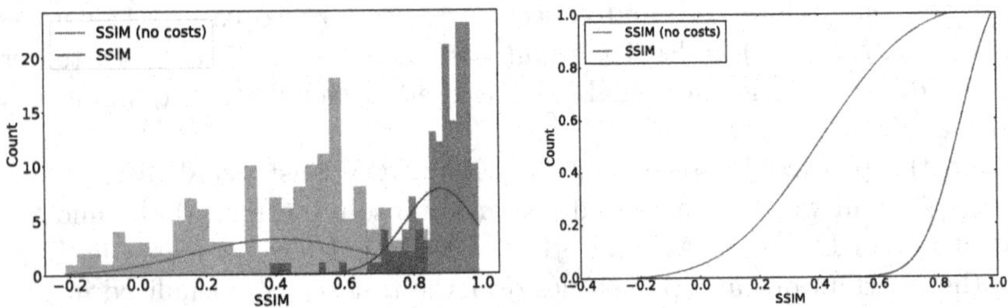

Fig. 6. SSIM measured over the testing set. Distribution (left) and cumulative probability (right).

$$\text{GDT\_HA} = \frac{p_{<0.5\text{Å}} + p_{<1\text{Å}} + p_{<2\text{Å}} + p_{<4\text{Å}}}{4} \tag{7}$$

where $p_{<n\text{Å}}$ indicates the percentage of an amino acid's coordinates in P whose distance from the corresponding amino acid's coordinates in T is below n Å.

Results for selected proteins are shown in Table 2. Note how both the GDT_TS and the GDT_HA generally increase by at least a factor of 2 when adding cost maps as an additional feature. Examples of the predicted coordinates and relative distance maps are given in Figs. 7, 8, 9 and 10.

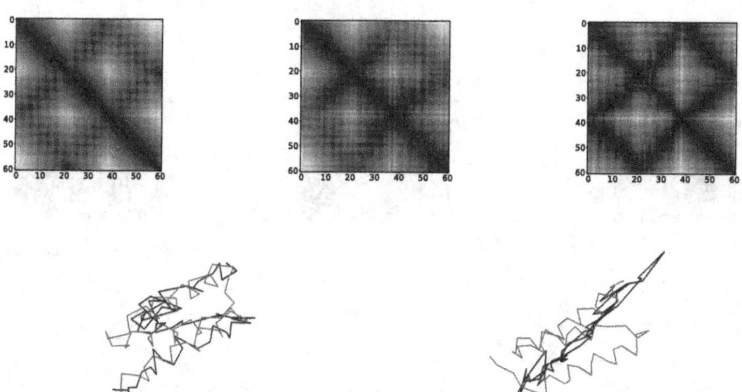

Fig. 7. Top row, from left to right: original, predicted and predicted (without cost maps) distogram for protein 2gomA. Bottom row: original (red) and predicted (blue) C_α coordinates. Left: with costs, right: without costs (Color figure online)

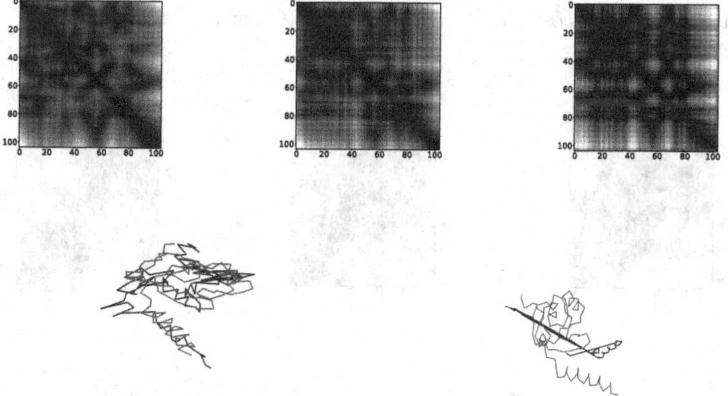

Fig. 8. Top row, from left to right: original, predicted and predicted (without cost maps) distogram for protein 1dm9A. Bottom row: original (red) and predicted (blue) C_α coordinates. Left: with costs, right: without costs (Color figure online)

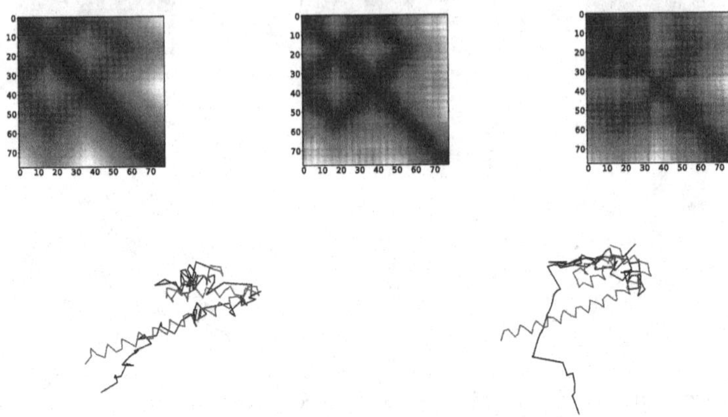

Fig. 9. Top row, from left to right: original, predicted and predicted (without cost maps) distogram for protein 2fztA. Bottom row: original (red) and predicted (blue) C_α coordinates. Left: with costs, right: without costs (Color figure online)

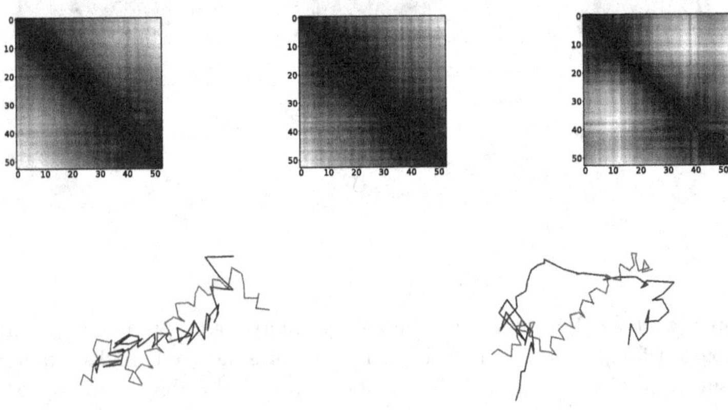

Fig. 10. Top row, from left to right: original, predicted and predicted (without cost maps) distogram for protein 2fyuK. Bottom row: original (red) and predicted (blue) C_α coordinates. Left: with costs, right: without costs (Color figure online)

Table 2. Metrics between original and predicted coordinates with and without (-) cost maps after SVD alignment. MAE and SSIM are measured on distance maps.

Protein	MAE	SSIM	GDT_TS	GDT_HA
2gomA	2.49	0.86	31.2	9.84
2gomA (-)	5.31	0.53	13.9	4.10
1zv1A	2.83	0.96	28.8	10.2
1zv1A (-)	4.78	0.46	14.0	3.39
1dm9A	3.58	0.95	25.5	7.93
1dm9A (-)	6.94	0.36	6.25	0.48
1m8nA	3.59	0.92	22.0	7.92
1m8nA (-)	5.04	0.49	11.0	2.08
2fztA	3.65	0.96	18.6	4.49
2fztA (-)	6.81	0.55	8.96	1.60
2fyuK	3.44	0.98	15.1	3.30
2fyuK (-)	6.69	0.82	4.72	1.41

5 Conclusions

In this paper we introduced a measure of the orientation between amino acids based on GA. We presented the ideas behind the modelling of a protein as a collection of planes, we introduced a measure of the "distance" between each pair of planes and arranged it in matrix form, i.e. a cost map.

We then employed these matrices as an additional feature in a GT + 3D projector pipeline to predict 3D coordinates of C_α atoms in proteins. We did so by adapting in graph form a dataset comprising several biochemical features already available in the literature, to which we added cost maps. Eventually, we compared the 3D coordinates predicted including cost maps with coordinates predicted without them.

We showed that our GA-based cost maps aids the convergence of the model and the prediction of more accurate coordinates in terms of GDT_TS and GDT_HA scores with respect to ground truth. In addition, the distance maps constructed from the coordinates predicted including costs are closer to the original distance maps in terms of both MAE and SSIM.

Despite training the model on a dataset of only 200 short proteins and for few iterations, we managed to obtain reasonable protein structures. We are confident that including cost maps on a larger scale problem (e.g. larger training set, more learning iterations, higher dimensionality of node and edge embeddings, etc.) can constitute an asset in PSP by increasing prediction accuracy with a minimal amount of additional information.

References

1. Jumper, J., et al.: Highly accurate protein structure prediction with AlphaFold. Nature **596**(7873), 583–589 (2021)
2. Thornton, J.M., Laskowski, R.A., Borkakoti, N.: AlphaFold heralds a data-driven revolution in biology and medicine. Nat. Med. **27**(10), 1666–1669 (2021)
3. Perrakis, A., Sixma, T.K.: AI revolutions in biology: the joys and perils of AlphaFold. EMBO Rep. **22**(11), e54046 (2021)
4. Torrisi, M., Pollastri, G., Le, Q.: Deep learning methods in protein structure prediction. Comput. Struct. Biotechnol. J. **18**, 1301–1310 (2020)
5. Kandathil, S.M., Greener, J.G., Jones, D.T.: Recent developments in deep learning applied to protein structure prediction. Proteins Struct. Funct. Bioinform. **87**(12), 1179–1189 (2019)
6. Pakhrin, S.C., Shrestha, B., Adhikari, B., Kc, D.B.: Deep learning-based advances in protein structure prediction. Int. J. Mol. Sci. **22**(11), 5553 (2021)
7. Baek, M., et al.: Accurate prediction of protein structures and interactions using a three-track neural network. Science **373**(6557), 871–876 (2021)
8. Jaderberg, M., Simonyan, K., Zisserman, A.: Spatial transformer networks. Adv. Neural Inf. Process. Syst. **28** (2015)
9. Li, N., Liu, S., Liu, Y., Zhao, S., Liu, M.: Neural speech synthesis with transformer network. In: Proceedings of the AAAI Conference on Artificial Intelligence, vol. 33, no. 01, pp. 6706–6713 (2019)
10. Kim, S., Lin, S., Jeon, S.R., Min, D., Sohn, K.: Recurrent transformer networks for semantic correspondence. Adv. Neural Inf. Process. Syst. **31** (2018)
11. Giuliari, F., Hasan, I., Cristani, M., Galasso, F.: Transformer networks for trajectory forecasting. In: 2020 25th International Conference on Pattern Recognition (ICPR), pp. 10335–10342. IEEE (2021)
12. Costa, A., Ponnapati, M., Jacobson, J.M., Chatterjee, P.: Distillation of MSA embeddings to folded protein structures with graph transformers. bioRxiv (2021)
13. Yang, J., Anishchenko, I., Park, H., Peng, Z., Ovchinnikov, S., Baker, D.: Improved protein structure prediction using predicted interresidue orientations. Proc. Natl. Acad. Sci. **117**(3), 1496–1503 (2020)
14. Adhikari, B.: A fully open-source framework for deep learning protein real-valued distances. Sci. Rep. **10**(1), 1–10 (2020)
15. Doran, C., Gullans, S.R., Lasenby, A., Lasenby, J., Fitzgerald, W.: Geometric Algebra for Physicists. Cambridge University Press, Cambridge (2003)
16. Dorst, L., Doran, C., Lasenby, J. (eds.): Springer, Heidelberg (2012). https://doi.org/10.1007/978-1-4612-0089-5
17. Lavor, C., Alves, R.: Oriented conformal geometric algebra and the molecular distance geometry problem. Adv. Appl. Clifford Algebras **29**(1), 1–15 (2019)
18. Alves, R., Lavor, C.: Geometric algebra to model uncertainties in the discretizable molecular distance geometry problem. Adv. Appl. Clifford Algebras **27**(1), 439–452 (2017)
19. Jumper, J., et al.: AlphaFold 2 (2020)
20. Lasenby, J., Hadfield, H., Lasenby, A.: Calculating the rotor between conformal objects. Adv. Appl. Clifford Algebras **29**(5), 1–9 (2019)
21. Eide, E.R.: Master's degree thesis. University of Cambridge, Camera Calibration using Conformal Geometric Algebra (2018)
22. Yun, S., Jeong, M., Kim, R., Kang, J., Kim, H.J.: Graph transformer networks. Adv. Neural Inf. Process. Syst. **32** (2019)

23. Hadfield H., Wieser E., Arsenovic A., Kern R.: The Pygae Team. Pygae/Clifford (2020)
24. Burley, S.K., Berman, H.M., Kleywegt, G.J., Markley, J.L., Nakamura, H., Velankar, S.: Protein Data bank (PDB): the single global macromolecular structure archive. Protein Crystallogr. 627–641 (2017)

How Does Geometric Algebra Support Digital Twin—A Case Study with the Passive Infrared Sensor Scene

Yilei Yin[1,2], Binghuang Pan[1,2], Chunye Zhou[1,2], Wen Luo[1,2(✉)], Zhaoyuan Yu[1,2], and Linwang Yuan[1,2]

[1] Key Laboratory of Virtual Geographic Environment, Nanjing Normal University, Ministry of Education, Nanjing 210023, China
luow1987@163.com

[2] Jiangsu Center for Collaborative Innovation in Geographical Information Resource Development and Application, Nanjing 210023, China

Abstract. Digital twin (DT) has been applied to increasingly complex systems, including environments, energy, and digital cities, due to advancement of data collecting, high-speed networks, big data, artificial intelligence, and other technologies. Because of the complexity of the actual world and newly suggested criteria for the construction of the linkage between the real and virtual spatial, developing and using DT has been significantly hampered. The classic modeling approaches of separating expression from analysis have grown to be a significant barrier. These problems can be resolved due to the benefits of geometric algebra (GA) expression and computation in multidimensional space. The paper studies the concept of DT, employs the essential principles of GA as a tool, and proposes the DT's modeling and analysis methods. To investigate how DT is formulated and built, a typical multi-factor coupling scenario of passive infrared sensor (PIR) was presented as an example. The results demonstrate the effectiveness of the approach presented in this paper in simulating human-sensor interactions, producing reaction records in real space, and successfully deriving the pedestrian trajectory from PIR recordings. The study presented in this article offers fresh perspectives on how to build DT in complicated scenarios and may also shed new light on how to analyze human behavior using PIR.

Keywords: Digital twin · Geometric algebra · Passive infrared sensors · Trajectory extraction

1 Introduction

Digital twin (DT), also known as digital mirroring, was first proposed by Professor Michael Grieves [1]. The fundamental concept is to employ information

Supported by the National Natural Science Foundation of China (No. 42130103 and 41601417).

E. Hitzer et al. (Eds.): ENGAGE 2022, LNCS 13862, pp. 96–108, 2023.
https://doi.org/10.1007/978-3-031-30923-6_8

technology to model physical entities so that they can interact with digital models of their properties and actions [2,3]. Because of their connection, homogeneity, modularity, and intelligence, DTs are not just mirrors of physical entities but also information receivers and feedbackers [4–6]. They may even act as prophets and foretellers of the actual world [7,8]

Real space, virtual space, data flow from real space to virtual space, and information flow from virtual space to real space are the four fundamental components of DT [3,9]. Real space and virtual space are fairly obvious among them, although data flow and information flow are often disregarded and even ignored as the link between real space and virtual space. This naturally leads to the current study on DTs concentrating on two primary areas: 1) The real-space detecting apparatus 2) Virtual space analysis and mining techniques [10]. It is challenging to build a connection between the two spaces since the two tasks are often independent and performed by different groups [11,12].

The separation between detecting and analysis is mostly seen in two aspects. The first is the variation in the underlying mathematical underpinning. The detecting and modeling techniques based on Euclidean geometry and computational geometry are still dominant since the representation of current DTs is more concentrated on the description of entities' geometries and structures. For the analysis model, the statistical methods based on the detecting data and the dynamic model based on the mechanism of system are more popular. There is still few solution to the contradiction between the geometric representation of things and the algebraic computation of their properties. Secondly, there is the inconsistency between the expression unit and the calculation unit, that is, entity modeling only solves the collection and expression of data, what's worse, analysis models are often limited by a specific field, which make a big inconsistency of these two jobs in the granularity and semantics. Therefore, in view of the difficulty of establishing the relationship between real space and virtual space, building the underlying theory that combines geometry and algebra, expression and calculation is crucial for the future application of DTs [12].

Geometric algebra (GA), also known as Clifford algebra [13], is a combination algebra based on dimensional operations. It is an algebraic language for describing and computing geometric problems, created based on the Hamilton quaternion and Grassmann's extended algebra [14]. Geometries can be represented and produced algebraically in a unified way in terms of dimensions and relationships based on the expression theory of GA. The inner product, outer product, and geometric product provide the fundamental strategies for the unified computation of geometric patterns and spatial topological relations in multidimensional and coordinate-independent environments [15]. GA is currently the theoretical foundation and computing tool for mathematical analysis, geographic information science, geometry, and other fields [16]. The aforementioned accumulation can serve as a theoretical and methodological foundation for the creation and evaluation of DTs.

Passive infrared sensors (PIR) is a kind of sensor based on the principle of pyroelectric effect, which detects infrared photoelectrons radiated from objects.

Typically, it is used in motion detectors for security alerts and behavioral study of people [17,18]. In the analysis of sensor response and human behavior, the established information mirror model of DT offers fresh perspectives for researching PIR. Scalability and verification capabilities of the DT-oriented PIR virtual scene allow for the construction and verification of analytical models of human behavior characteristics. Due to the complexity of real space, the PIR scene must take into account not only the connectivity between various sensor nodes and their spatial-temporal responses to pedestrians in the scene, but also the integrated expression of the sensor topology network that is produced by the connectivity of the sensor nodes and the pedestrians' trajectories [19]. It is challenging to properly characterize the spatial-temporal link between sensor nodes and sensor response using traditional methods, which has constraints in the representation of high-dimensional relationships and dynamic trajectories.

This paper studied the concept of DT and used geometric algebra to propose the modeling and analysis methods of DT. The expression and construction of DT were investigated using the PIR scene as an example. The viability of the strategy described in this research was confirmed using the PIR simulation scenario that was built to study sensor response and human behavior. The work was divided into three sections: Sect. 2, which covered the theoretical underpinnings and fundamental concepts, Sect. 3, which presented the building of GA-based DT. Then, in part 4, DT modeling of the PIR scene was shown. In Sect. 5, the conclusion and debates were presented.

2 Basic Idea

The study adds a mathematical space that facilitates data modeling and information simulation to Professor Grieves' DT expression model from a realizable and calculable standpoint, as seen in Fig. 1. The definition of algebraic system is related to the abstract mode of the real world, solving practical problems through the definition of dimensions and metrics. To achieve the unified representation of objects in the complex real space, object expression requires the usage of GA's multivector structure. The GA operator realizes the object's function, and the operator must be compatible with the object's real-space objective law. After that, it will be possible to build real-world analysis and mining methods based on GA equations and use the information from virtual space to inform real-world planning and design.

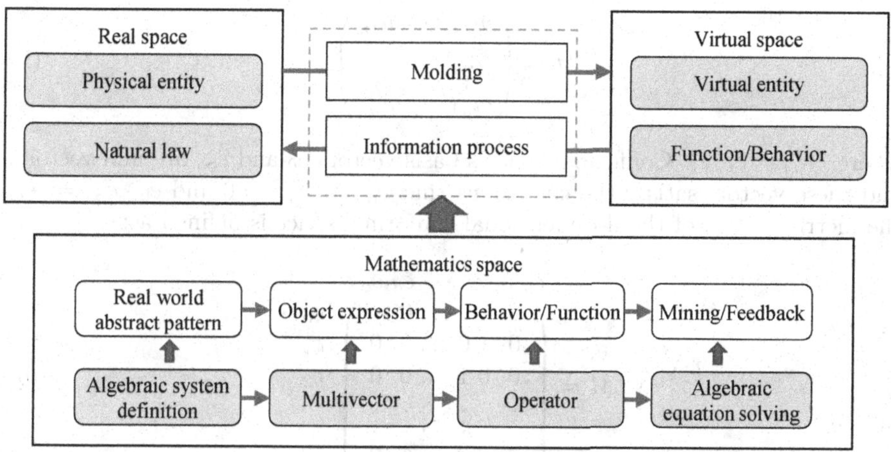

Fig. 1. Research idea and Conceptual diagram.

3 Construction of Geometric Algebra-based Digital Twin

GA is used to explore scene representation and computation in order to advance the development of DT's capacity to be computed, applied, and analyzed. A consistent multi-element modeling framework is created for DT objects in various dimensions and kinds. The framework makes it possible to represent the function and behavior of physical entities in real space.

3.1 Geometric Algebraic Systems Definition

GA defines basis vectors which are linearly independent to represent space and uses blades to represent fundamental objects. To enhance the blade's expressive capacity, more dimensions should always be added [20]. As an example, the Euclidean space R^3 for a three-dimensional space can be created by specifying a set of base vectors $\{e_1, e_2, e_3\}$ or by adding a dimension e_0 to create a homogenous space A^3. Homogeneous space, as opposed to Euclidean space, allows for the representation of flat objects like lines and planes. Similar to this, the conformal space C^3 is built by adding the extra dimensions e_0 and e_∞. The conformal space, as opposed to the homogeneous space, allows for the representation of rounded objects like spheres and circles. The conformal geometric algebraic (CGA) expands the dimension by adding positive space e_+ and negative space e_-, and via a sequence of transformations turns it into a parabolic space made up of e_0 and e_∞ [21].

The distinction between positive and negative space demonstrates the various operational properties of the basis vectors in GA. The following definition of a metric matrix of the space's basis can be used to describe this feature:

$$M = \begin{pmatrix} m_{11} & \cdots & m_{1n} \\ \vdots & \ddots & \vdots \\ m_{n1} & \cdots & m_{nn} \end{pmatrix} \tag{1}$$

where, $m_{ij} = e_i \cdot e_j$. Conformal space's basis vectors e_0 and e_∞ are null vectors, and these vectors satisfy the conditions that $e_0^2 = 0, e_\infty^2 = 0$ and $e_0 \cdot e_\infty = -1$, the metric matrix of the d-dimensional conformal space is defined as:

$$M = \begin{array}{c} \begin{matrix} e_0 & e_1 & e_2 & \cdots & e_d & e_\infty \end{matrix} \\ \begin{pmatrix} 0 & 0 & 0 & \ldots & 0 & -1 \\ 0 & 1 & 0 & \ldots & 0 & 0 \\ 0 & 0 & 1 & \ldots & 0 & 0 \\ \vdots & \vdots & \vdots & \ddots & \vdots & \vdots \\ 0 & 0 & 0 & \cdots & 1 & 0 \\ -1 & 0 & 0 & \cdots & 0 & 0 \end{pmatrix} \end{array} \begin{matrix} e_0 \\ e_1 \\ e_2 \\ \vdots \\ e_d \\ e_\infty \end{matrix} \tag{2}$$

3.2 Object Representation Based on Geometric Algebra

Blades, which expand dimensions via outer products, are the foundation for the representation of objects in GA. k-blade is obtained by the outer product of k linearly independent vectors. In CGA, the outer product inherits the spatial structure (Grassmann structure) in the geometry construction. Therefore, multi-dimensional geometric objects can be expressed uniformly under the framework of CGA [16,22]. The Fig. 2 below shows the representation of basic geometric objects in CGA.

3.3 Geometric Algebra Operator Definition

An essential starting point for the analysis and mining of object characteristics in virtual space is the representation of entity functions and behaviors as operators in real space. The aforementioned GA definition can be used to establish the dimensional unity and object-independent operation of geometric objects, resulting in a unified interface for the construction of computing operators. The GA operator library is created using the basic operators of GA, which are primarily divided into three categories: dimension operators for object building, transformation operators for object transformation, and relation operators for calculating object relationships.

Dimension operators are used to construct spatial objects, which are organized and decomposed by adding or subtracting dimensions. The majority of dimensional operations are binary operations with data parameters $ParD$ as their input and output objects.

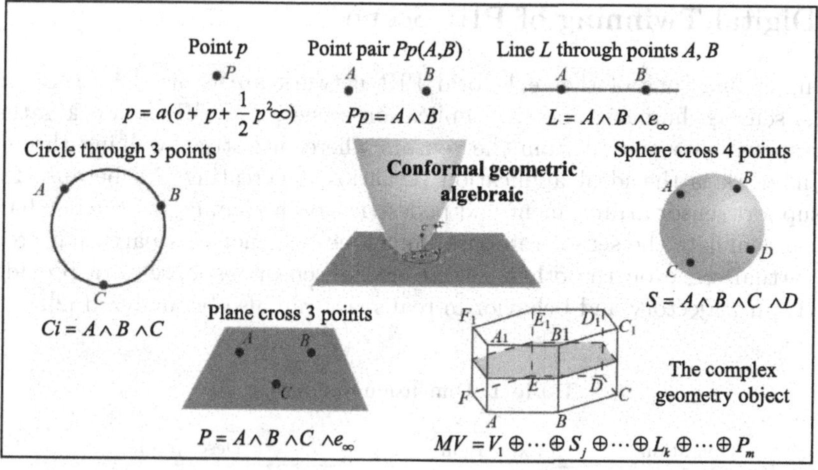

Fig. 2. Representation of Basic Shapes in Conformal Geometry Algebraic.

$$f_{op_d} = f(2, \{ParD_1, ParD_2\}, op_d) = ParD_3 \tag{3}$$

Table 1 demonstrates that dimensional operators mostly consist of GA fundamental operators like inner, outer, and geometric products as well as dimensional operators like projection and reflection.

Transformation operators are used to change the algebraic form and spatial attribute of the parameters. It has the ability to alter spatial aspects as well as optimize the algebraic structure (such as translation and rotation). The transformation operator has both unary and binary forms.

$$\begin{cases} f_{op_c} = f(1, \{Par_1\}, op_c) = Par_2 \\ f_{op_c} = f(2, \{ParD_1, ParT_1\}, op_c) = ParD_2 \end{cases} \tag{4}$$

where Par_i is the general parameter, $ParT_i$ and $ParD_i$ represent transformation parameters and data parameters, respectively. Common transformation operators are shown in Table 2.

Relational operators, which are primarily used to determine the connection between objects, are based on subspace calculation operators. Because of this, all relational operators do binary computations, and all of their results are semantic parameters.

$$f_{op_r} = f(2, \{ParD_1, ParD_2\}, op_r) = ParS_1 \tag{5}$$

where $ParS_i$ is the semantic parameter. The relational operators are shown in Table 3.

4 Digital Twinning of PIR Scene

The main functions of the real-world PIR network are realized by three elements: sensors, humans, and the application scenes. The PIR network gathers the response information from the sensors where pedestrian walking through. PIR network is the ideal application scenarios for creating DT because they can support sensor arrangement and pedestrian path planning on the one hand, and can simulate the sensor response under the behavior of a particular crowd in a virtual scene on the other. Based on the sensor response data provided, pedestrian trajectory and behavior in real space can also be analyzed [2].

Table 1. Dimension operators

Type	Operation	Expression	Description
Basic operators	Outer product	$\mathrm{op}(a,b) = a \wedge b$	Basic calculation of dimensions. op() and ip() are used to increase and decrease dimensions; gp() and iv() can generate multidimensional objects.
	Inner product	$\mathrm{ip}(a,b) = a \cdot b$	
	Geometric product	$\mathrm{gp}(a,b) = a \cdot b + a \wedge b$	
	Invert	$\mathrm{iv}(a) = a^{-1} = \mathrm{rv}(a)/(a{*}\mathrm{rv}(a))$	
Dimension operators	Extract grade i	$\mathrm{grd}(a,i) = \langle a \rangle_i$	Dimension extraction operators, which can be used to extract specific subspaces
	Norm	$\mathrm{norm2}(a) = a * \mathrm{rv}(a)$	
	Duality	$\mathrm{dual}(a) = a^* = a/I_m$	
	Projection	$\mathrm{Prj}(a,b) = (a \cdot b)b^{-1}$	
	Reflection	$\mathrm{Rej}(a,b) = bab^{-1}$	

rv(a) is the reverse operation, which will be described in Table 2.

Table 2. Transformation operators

Type	Operation	Expression	Description
Sequence adjustment operators	Reverse	$\mathrm{rv}(a) = (-1)^{n(n-1)/2}a$	The order adjustment operators of dimension components
	Grade involution	$\mathrm{giv}(a) = (-1)^n a$	
	Conjugate	$\mathrm{con}(a) = (-1)^{n(n+1)/2}a$	
Transformation operators	Reflect	$\mathrm{Ref}(a, D) = (-1)^{nd} DaD^{-1}$	The reflection of a on D, where d is the dimension of D
	Scale	$\mathrm{Scal}(\rho) = (1 + e_\infty)\rho + (1 - e_\infty)\rho^{-1} = e^{e_\infty \ln \rho}$	Scale at rate ρ
	Translate	$\mathrm{Trans}(t) = 1 + \frac{1}{2}te_\infty = e^{-\frac{t}{2}e_\infty}$	Translation distance t
	Rotor	$\mathrm{Rotor}(\theta, l) = \cos(\frac{\theta}{2}) - \sin(\frac{\theta}{2})l = e^{-\frac{\theta}{2}l}$	Rotation angle θ around axis l

Table 3. Spatial relation calculation operators.

Type	Operation	Expression	Description
Relational measure operators	point-point	$dst\_pt(A, B) = \sqrt{-2A' \cdot B'}$	The distance from point A to B.
	point-line	$dst\_pl(P, l_{AB}) = (e_\infty \wedge A' \wedge B' \wedge P')^*$	
	point-circle	$dst\_ps(A, S_{BCD}) = \frac{A' \wedge B' \wedge C' \wedge D'}{e_\infty \wedge A' \wedge B' \wedge C'}$	
	line-circle	$dst\_ls(l_{AB}, S_{CDE}) = ((e_\infty \wedge A' \wedge B') \cap (C' \wedge D' \wedge E'))^2$	The distance between two objects reflected by the size and the positive and negative values.
	circle-circle	$dst\_ss(S_{ABC}, S_{DEF}) = ((A' \wedge B' \wedge C') \cap (D' \wedge E' \wedge F'))^2$	
Topological relation judgment operators	intersection	$meet(A, B) = A \cap B = B^* \cdot A$	The intersection of A and B
	union	$join(A, B) = A \cup B = A \wedge (M^{-1} \cdot B)$	Minimum computing space, M is the largest common divisor of A and B

A' and B' are the CGA expression of point A and B.

4.1 Data and Framework

The sensor response data from Mitsubishi Electric Research Labs (MERL), as well as the spatial layout data and sensor network data of the sensor network entity scene, were utilized as the data sources for this paper. Between March 21, 2006, and May 24, 2007, 156 sensors' reaction data to human movement were gathered in the dataset.

Figure 3 illustrates the DT modeling framework. Using GA, we can establish the virtual spaces V_1 for expression, V_2 for PIR scene interaction response, and V_3 for pedestrian trajectory mining. The specification of the GA system, object representation, and functional operator of each virtual space are distinct as a result of the various application aims. In order to achieve the generation of the sensor response record when the pedestrian moves in the scene, the virtual space V_2 needs to define multivector representations of the buildings, sensors, and people in C3GA. It is also necessary to construct the response operator of the sensor and the motion operator of the pedestrian. To create the whole sensor network scenario, virtual space V_3 must specify network nodes and network edges in NnGA (nD network algebra [23]). The network extension and trajectory extension operators can be functionally developed in order to determine the actual trajectories of pedestrians using sensor data that has already been collected.

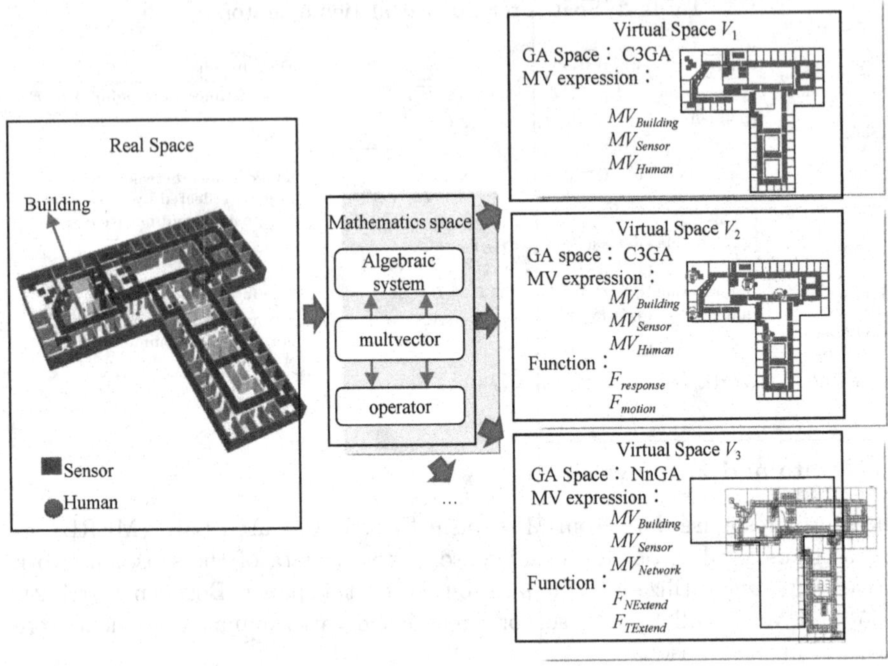

Fig. 3. Digital Twin Modeling Framework.

4.2 Digital Twinning for Interaction of PIR scene

Based on the above framework, we established the sensor-human response virtual scene in conformal space (Fig. 4). In the response process, the individual human under the action of motion F_{motion} continuously interacts with the response operator $F_{response}$ of the sensors. The sensors' 0–1 response sequence, X^k_{sensor}, was computed to provide response data. In the real world, crowds moving in the same location will also provide a series of response data. In order to evaluate the accuracy of the reaction simulation in virtual scene, MERL's event log states that a sample of the sensor response data collected during the fire evacuation was utilized for comparison. Figure 4 depicts the actual response sequence in the real world and the simulated response sequence in the virtual scene. Since the actual human movement is unknown, there is a slight difference between the simulated response results of the sensor in the virtual and real scene. The main gaps are found in stairwells, elevator halls, and other areas where people move frequently.

The correlation coefficient between the number of responses in the actual world and the virtual scene is 58.36%. In general, the virtual world can accurately simulate the human-sensor reaction if the challenging aspects, including randomness and unpredictability in the movement of the crowd, are removed. The accuracy rating for our simulation of the sensor response during the early stage of evacuation was 67.62%. This could be the case because human behavior is more irregular during evacuation, making it challenging to simulate using the usual human behavior pattern. The key to increasing simulation accuracy is choosing the correct behavior model to build F_{Human}.

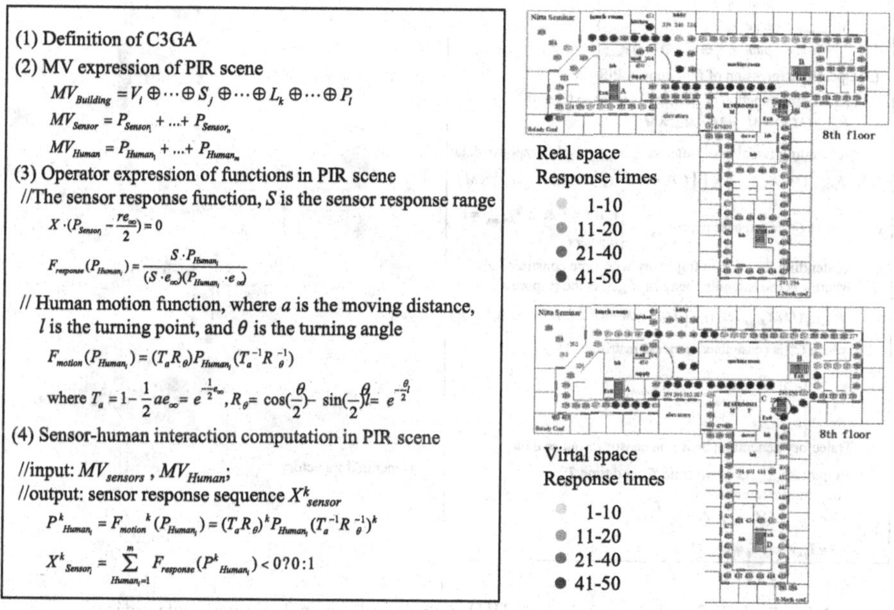

Fig. 4. DT for data generating of sensor-human response.

4.3 Digital Twinning for Data Mining of PIR Scene

On the basis of the PIR network model and response data in real space (Fig. 5), we also created a mining virtual scene to analyze pedestrians' trajectories. The network space NnGA was established because the pedestrians' trajectories are constrained to the network of actual space. Additionally, the nearby sensor nodes will respond continually in a finite amount of time when human passes by. Although the PIR cannot directly detect the trajectory, it is feasible to mine potential trajectories based on the spatiotemporal correlation of the responses of Sufficient number of sensors nearby. We developed trajectory extending algorithms and trajectory extending algorithms under time constraints based on this characteristic (the left part of Fig. 5). The trajectories can then be extracted from the response data (the right part of Fig. 5).

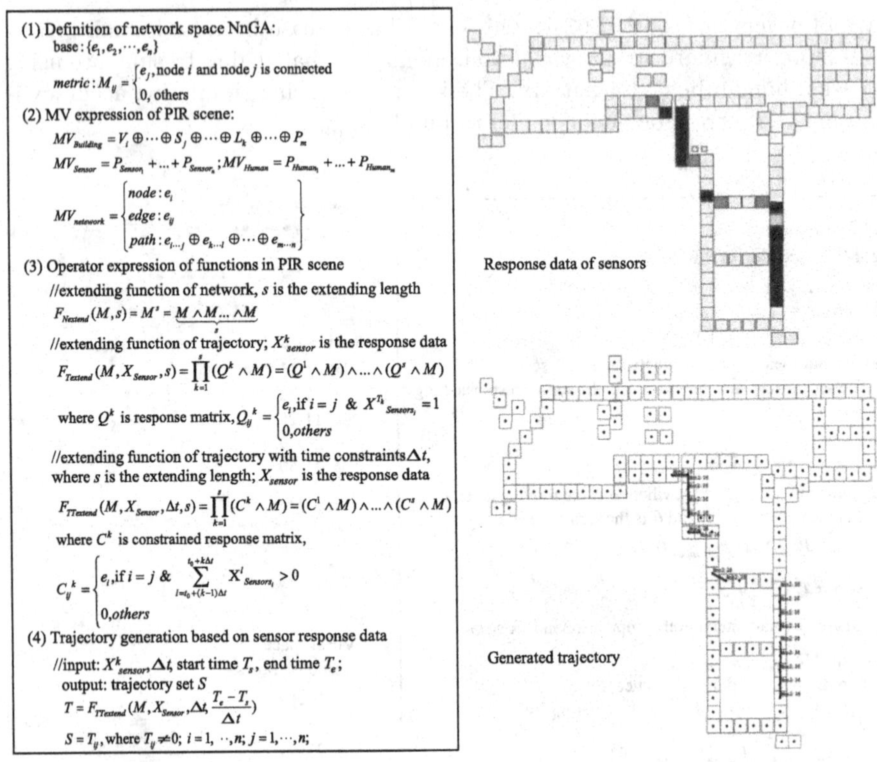

Fig. 5. DT for data mining of PIR scene and trajectory reconstruction.

5 Conclusion and Discussion

This paper proposes a DT modeling and analysis approach based on GA considering the absence of DT's underlying mathematical theory. GA's space definability and expression unity can be used to construct DT systems for a variety of purposes. The behavior and function of entities in virtual space can be defined, and the relationship and feedback between virtual space and actual space can be achieved, by building corresponding operators in GA.

The paper proposes a framework for DT modeling based on GA, there is still a lot of work to be carried out in the future. Include as follows: 1) Facing the complex real space, it is necessary to construct a richer GA system, especially in the expression of semantics and knowledge; 2) Basic geographic transformation, metric, and relational computing operators are provided by GA, but they often need to be expanded for real-world issues. It is essential to provide a practical and expandable GA operator library; 3) While space definability and operator extensibility are desirable characteristics of GA, they also place extra expectations on DT developers. Building a meta-space and meta-operator library will therefore be a crucial area of study for universal DT modeling.

References

1. Grieves, M.: Product lifecycle management: the new paradigm for enterprises. Int. J. Prod. Dev. **2**(1/2), 71–87 (2005)
2. Chen, Y., Shen, W., Wang, X.: The internet of things in manufacturing: key issues and potential applications. IEEE Syst. Man Cybern. Mag. **4**(1), 6–15 (2018)
3. Tao, F., Zhang, H., Liu, A., Nee, A.Y.C.: Digital twin in industry: state-of-the-art. IEEE Trans. Industr. Inf. **15**(4), 2405–2415 (2019)
4. Fertig, A., Weigold, M., Chen, Y.: Machine learning based quality prediction for milling processes using internal machine tool data. Adv. Industr. Manuf. Eng. **4**, 100074 (2022)
5. Vandermerwe, S., Rada, J.: Servitization of business: adding value by adding services. Eur. Manag. J. **6**(4), 314–324 (1988)
6. Bumann, A.: Navigating the black box: generativity and incongruences in digital innovation. Ph.D. thesis, Chalmers University of Technology (2022)
7. Hallerbach, S., Xia, Y., Eberle, U., Koester, F.: Simulation-based identification of critical scenarios for cooperative and automated vehicles. SAE Int. J. Connect. Autom. Veh. **1**(2), 2018-01-1066 (2018)
8. Guanghui, Z., Chao, Z., Zhi, L., Kai, D., Chuang, W.: Knowledge-driven digital twin manufacturing cell towards intelligent manufacturing. Int. J. Prod. Res. **58**, 1034–1051 (2020)
9. Grieves, M., Vickers, J.: Digital twin: mitigating unpredictable, undesirable emergent behavior in complex systems. In: Kahlen, F.-J., Flumerfelt, S., Alves, A. (eds.) Transdisciplinary Perspectives on Complex Systems, vol. 1, pp. 85–113. Springer, Cham (2017). https://doi.org/10.1007/978-3-319-38756-7_4
10. Shirowzhan, S., Tan, W., Sepasgozar, S.: Digital twin and cybergis for improving connectivity and measuring the impact of infrastructure construction planning in smart cities. ISPRS Int. J. Geo Inf. **9**, 240 (2020)
11. Qi, Q., et al.: Enabling technologies and tools for digital twin. J. Manuf. Syst. **58**, 3–21 (2019)
12. Jones, D., Snider, C., Nassehi, A., Yon, J., Hicks, B.: Characterising the digital twin: a systematic literature review. CIRP J. Manuf. Sci. Technol. **29**, 36–52 (2020)
13. Clifford, W.K.: Applications of Grassmann's extensive algebra. Am. J. Math. **1**(4), 350–358 (1878)
14. Yuan, L., Lü, G., Luo, W., Yu, Z., Sheng, Y.: Geometric algebra method for multidimensionally-unified GIS computation. Chin. Sci. Bull. **57**(7), 802–811 (2012)
15. Yu, Z., Luo, W., Yuan, L., Hu, Y., Zhu, A.: Geometric algebra model for geometry-oriented topological relation computation. Trans. GIS **20**, 259–279 (2016)
16. Yuan, L., Yu, Z., Luo, W., Lin, Y.: Geometric algebra for multidimension-unified geographical information system. Adv. Appl. Clifford Algebras **23**(2), 497–518 (2013)
17. Jin, X., Sarkar, S., Ray, A., Gupta, S., Damarla, T.: Target detection and classification using seismic and PIR sensors. IEEE Sens. J. **12**(6), 1709–1718 (2012)
18. Yang, D., Xu, B., Rao, K., Sheng, W.: Passive infrared PIR-based indoor position tracking for smart homes using accessibility maps and a-star algorithm. Sensors **18**, 332 (2018)
19. Xiao, S., Yuan, L., Luo, W., Li, D., Zhou, C., Yu, Z.: Recovering human motion patterns from passive infrared sensors: a geometric-algebra based generation-template-matching approach. ISPRS Int. J. Geo Inf. **8**(12), 1–19 (2019)

20. Dorst, L., Fontijne, D., Mann, S.: Geometric Algebra for Computer Science - An Object-Oriented Approach to Geometry. In The Morgan Kaufmann Series in Computer Graphics (2007)
21. Perwass, C.: Geometric Algebra with Applications in Engineering. Springer, Heidelberg (2009). https://doi.org/10.1007/978-3-540-89068-3
22. Dorst, L., Lasenby, J.: Guide to Geometric Algebra in Practice. Springer, London (2011). https://doi.org/10.1007/978-0-85729-811-9
23. Yuan, L., Yu, Z., Luo, W., Zhang, J., Hu, Y.: Clifford algebra method for network expression, computation, and algorithm construction. Math. Methods Appl. Sci. **37**(10), 1428–1435 (2014)

Signal Processing with Octonions

Beurling's Theorem Associated with Octonion Algebra Valued Signals

Youssef El Haoui[1(✉)] and Mohra Zayed[2]

[1] Ecole Normale Supérieure, Moulay Ismail University of Meknès, Meknes, Morocco
y.elhaoui@umi.ac.ma

[2] Mathematics Department, College of Science, King Khalid University, Abha, Saudi Arabia
mzayed@kku.edu.sa

Abstract. The octonion Fourier transform (OFT) is a hypercomplex Fourier transform that extends the quaternion Fourier transform. This paper deals with the generalization of Beurling's uncertainty principle for octonion-valued signals and on $\mathbb{R}^3$, and therefore extends three uncertainty principles (UP), namely Hardy's UP, Gelfand–Shilov's UP, and Cowling–Price's UP, to the OFT domain.

Keywords: Fourier transform · octonion algebra · Clifford algebra · uncertainty principle · Beurling's theorem

1 Introduction

The octonion algebra is an eight-dimensional (8D) algebra that extends the four-dimensional (4D) quaternion algebra and the complex and real number algebras. However, if we lose commutativity in the associative quaternion algebra, we lose both associativity and commutativity in the octonion algebra, which makes the calculations non-obvious.

During the last few years, a great deal of attention has been paid to the study of hyper-complex signals, including quaternionic, octonionic, and Clifford algebraic signals, in general. Hyper-complex Fourier analysis has found many practical uses, especially in color image processing. Within this context, researchers have brought to light octonion signals that generalize quaternion signals through the octonion Fourier transform (OFT). Recently, OFT has emerged as a research stream in the hyper-complex Fourier domain, and many results have been established for real and octonion-valued functions. For example, we refer to [3,4,10–12].

The uncertainty principle (UP) is a crucial tool in mathematics and physics, especially in quantum physics and signal processing. In quantum mechanics, the UP was first suggested by the German physicist W. Heisenberg in 1927. UP

This work is dedicated to Mohra Zayed's parents, who believed in their daughter and supported her. May God bless them.

E. Hitzer et al. (Eds.): ENGAGE 2022, LNCS 13862, pp. 111–122, 2023.
https://doi.org/10.1007/978-3-031-30923-6_9

roughly states that the more precisely the position of a particle is determined, the less precisely its momentum can be known, and vice versa. From the point of view of signal processing, UP has the following characterization: "It is impossible to accurately locate a signal simultaneously in the time domain and in the frequency domain". There are many forms of the uncertainty principle associated with the Fourier transform, such as the Heisenberg uncertainty principle, Hardy, Gelfand–Shilov, and Cowling–Price UPs. [6]. A more general version of the uncertainty principle is called Beurling's theorem, where decay has been measured in terms of a single integral estimate involving a signal f and its Fourier transform $\hat{f}$.

Theorem 1. *(Beurling uncertainty principle)*
Suppose that $f \in L^1(\mathbb{R}^n)$ is such that

$$\int_{\mathbb{R}^n}\int_{\mathbb{R}^n} |f(x)||\hat{f}(y)|e^{2\pi|x||y|}dxdy < \infty$$

where $\hat{f}(y) = \int_{\mathbb{R}^n} f(x)e^{-2\pi i\langle x,y\rangle}dx$, then $f = 0$ almost everywhere.

The original proof of Theorem 1 is based on results of complex analysis and, in particular, on the Phragmén-Lindelöf principle, and was given, without a found proof, for $n = 1$, by A. Beurling at the end of the 1980s,s, and proved after Hörmander [9] in 1991, then it was extended on $\mathbb{R}^n$, for $n \geq 1$ by by S.C. Bagchi et al. in [1]. The strength of Theorem 1 consists in its immediate implication of the weak form of Hardy's uncertainty principle [7],

Theorem 2. *(Hardy uncertainty principle)*
Let $f \in L^2(\mathbb{R}^n)$ satisfy $|f(x)| \leq Ce^{-\pi\alpha|x|^2}$, and $|\hat{f}(y)| \leq C'e^{-\pi\beta|y|^2}$, where C and C' are two positive constants.

(i) If $\alpha\beta > 1$, then $f = 0$ almost everywhere.
(ii) If $\alpha\beta = 1$, then $f(x) = ce^{-\pi\alpha|x|^2}$, where c is a positive constant.
(iii) else there are infinitely many linearly independent functions satisfying the conditions.

The purpose of this paper is to: Firstly, we establish a relationship between the octonion Fourier transform and the 3-dimension Clifford–Fourier transform, which we believe will be useful in the future for the mathematical community to establish new results in octonion analysis. This relationship is given also in terms of the octonion norm. Secondly, by reducing the calculations to real-valued functions on $\mathbb{R}^3$, using a new norm for OFT, we establish the Beurling uncertainty theorem for the first time in the OFT domain, which allows us, by following the classical approach, to derive three variants of UPs including Hardy's UP. The rest of the paper is structured as follows. Section 2 introduces the necessary background of octonion algebra over $\mathbb{R}^3$, and the Clifford algebra $C\ell_{0,3}$ and its associated Clifford-Fourier transform. Then Sect. 3 reviews the octonion Fourier transform (OFT) and states several of its important properties. Section 4 establishes the main results of the paper, i.e. Beurling's UP, Hardy's UP, Gelfand–Shilov's UP, and Cowling–Price's UP for the OFT.

2 On Octonion and Clifford Algebra $C\ell_{0,3}$

This section aims to provide a deeper understanding of the octonion algebra, which acts as a backbone for further developments. It also provides a definition of the Clifford algebra of signature $(0, 3)$ and recalls the associated three-dimensional (3D) Clifford-Fourier transform.

2.1 Octonion Algebra

The algebra of octonions $\mathbb{O}$ is is defined as a non-commutative and non-associative 8D real algebra over $\mathbb{R}^3$ with the basis $\{e_0, e_1, e_2, e_3, e_4, e_5, e_6, e_7\}$, where e_0 is the unit element 1 which is omitted whenever clear from the context. The octonion $o \in \mathbb{O}$ can be explicitly expressed as follows:

$$\begin{aligned} o &= o_0 + o_1e_1 + o_2e_2 + o_3e_3 + o_4e_4 + o_5e_5 + o_6e_6 + o_7e_7 \\ &= (o_0 + o_1e_1 + o_2e_2 + o_3e_3) + (o_4 + o_5e_1 + o_6e_2 + o_7e_3) \odot e_4 \\ &= a + b \odot e_4 \end{aligned} \tag{1}$$

where $\odot$ denotes the octonion multiplication, $o_0, \ldots, o_7 \in \mathbb{R}$, and a and $b \in \mathbb{H}$ are quaternions. The form (1) is called the *quaternion form* of an octonion.

Table 1. Octonions Multiplication Table

	Basis elements							
$\odot$	1	e_1	e_2	e_3	e_4	e_5	e_6	e_7
1	1	e_1	e_2	e_3	e_4	e_5	e_6	e_7
e_1	e_1	-1	e_3	$-e_2$	e_5	$-e_4$	$-e_7$	e_6
e_2	e_2	$-e_3$	-1	e_1	e_6	e_7	$-e_4$	$-e_5$
e_3	e_3	e_2	$-e_1$	-1	e_7	$-e_6$	e_5	$-e_4$
e_4	e_4	$-e_5$	$-e_6$	$-e_7$	-1	e_1	e_2	$-e_3$
e_5	e_5	e_4	$-e_7$	e_6	$-e_1$	-1	$-e_3$	e_2
e_6	e_6	e_7	e_4	$-e_5$	$-e_2$	e_3	-1	$-e_1$
e_7	e_7	$-e_6$	e_5	e_4	$-e_3$	$-e_2$	e_1	-1

Octonion algebra multiplication is given in Table 1, describing the results of multiplying the element in the ith row by the element in the jth column from the right. We easily observe that for $1 \leq i, j \leq 7$,

$$e_i \odot e_i = -1, \qquad \text{and} \qquad e_i \odot e_j = -e_j \odot e_i \quad \text{if } i \neq j.$$

In addition, in the table, it is easily observed that the multiplication is non-associative., from (e.g., $e_4 = e_2 \odot (e_3 \odot e_5) \neq (e_2 \odot e_3) \odot e_5 = -e_4$).

Moreover, by identifying, the algebras span$\{1\}$, span$\{1, e_4\}$, and span$\{1, e_1, e_2, e_3\}$, respectively, with the real numbers $\mathbb{R}$, the complex numbers $\mathbb{C}$, and the quaternion algebra $\mathbb{H}$, we remark that $\mathbb{R}, \mathbb{C}$, and $\mathbb{H}$ are a sub-algebras of $\mathbb{O}$.

Convention: with regard to the non-associative nature of octonion algebra, by convention, the multiplication of octonions throughout the article is from left to right, so that

$$p_1 \odot p_2 \odot \cdots \odot p_n = (\cdots(((p_1 \odot p_2) \odot p_3) \odot p_4) \cdots) \odot p_n$$

for any $p_i \in \mathbb{O}$.

We will call the part $Sc(o) := o_0$, the scalar part of o and $Vec(o) := o - o_0$, the vector part of o.
The octonion conjugate is defined by

$$\overline{o} = o_0 - Vec(o). \tag{2}$$

For all $o, p \in \mathbb{O}$, we have

$$\overline{\overline{o}} = o, \quad and \quad \overline{o \odot p} = \overline{p} \odot \overline{o}. \tag{3}$$

The norm of o equals

$$|o| = \sqrt{o \odot \overline{o}} = \sqrt{\sum_{i=0}^{i=7} o_i^2}. \tag{4}$$

Note that the octonion norm fulfils the law of composition, i.e., for any $o, p \in \mathbb{O}$,

$$|o \odot p| = |o|\ |p|, \tag{5}$$

then, we say then that the octonion algebra has a multiplicative norm.
The exponential of an octonion o is given by:

$$e^o = \sum_{i=0}^{\infty} \frac{o^i}{i!}, \tag{6}$$

where $o^i = \underbrace{o \odot o \odot o \cdots \odot o}_{i \ times}$. Since octonions are non-commutative, the relation $e^{o+p} = e^o \odot e^p$ will not always be true. However, this property is confirmed when o and p are commutative.

Lemma 1. *Let $\theta \in \mathbb{R}$ and $\mu \in \mathbb{O}$, with $\mu^2 = -1$. Then we have the following natural generalization of Euler's formula for octonion algebra*

$$\begin{aligned} e^{\theta\mu} &= \sum_{i=0}^{\infty} \frac{(\theta\mu)^i}{i!} = \sum_{i=0}^{\infty} (-1)^i \frac{\theta^{2i}}{(2i)!} + \mu \sum_{i=0}^{\infty} (-1)^i \frac{\theta^{2i+1}}{(2i+1)!} \\ &= \cos\theta + \mu \sin\theta \end{aligned} \tag{7}$$

Consequently, by remarking that $\frac{Vec(o)}{|Vec(o)|}$ is a square of -1, we get the following lemma

Lemma 2. *Every octonion $o \in \mathbb{O} - \{0\}$ can be written in polar form by*

$$o = |o|e^{\theta\mu} \tag{8}$$

where $\theta = \arctan\left(\left|\frac{Vec(o)}{Sc(o)}\right|\right)$ *and* $\mu = \frac{Vec(o)}{|Vec(o)|} = \frac{\sum_{i=1}^{i=7} o_i e_i}{\sqrt{\sum_{i=1}^{i=7} o_i^2}}$ *belongs to the unit sphere*

$$\mathcal{S}^2 := \{o \in \mathbb{O} : |o|^2 = 1\}$$

of the Euclidean space $\mathbb{R}^7$.

Lemma 3. *[10] Let $a, b \in \mathbb{H}$, then*
(i). $e_4 \odot a = \overline{a} \odot e_4$ (ii). $e_4 \odot (a \odot e_4) = -\overline{a}$ (iii). $(a \odot e_4) \odot e_4 = -a$ (iv). $a \odot (b \odot e_4) = (b \odot a) \odot e_4$(v). $(a \odot e_4) \odot b = (a \odot \overline{b}) \odot e_4$ (vi). $(a \odot e_4) \odot (b \odot e_4) = -\overline{b} \odot a$.

Furthermore, for an octonion $o = a + b \odot e_4,\ a, b \in \mathbb{H}$ in the quaternion form, we have [10, Lemma 2.11]

$$\overline{o} = \overline{a} - b \odot e_4. \tag{9}$$

and

$$|o|^2 = |a|^2 + |b|^2. \tag{10}$$

An octonion valued function $f : \mathbb{R}^3 \mapsto \mathbb{O}$ may be written as

$$f = f_0 + f_1 e_1 + f_2 e_2 + \cdots + f_7 e_7 \tag{11}$$

where each f_i is a real valued function.
For $1 \leq p < \infty$, we denote the spaces $L^p(\mathbb{R}^3, \mathbb{O})$ as the collection of all octonion valued functions $f : \mathbb{R}^3 \mapsto \mathbb{O}$ with the finite norm

$$\|f\|_p = \left(\int_{\mathbb{R}^3} |f(\boldsymbol{x})|^p d\boldsymbol{x}\right)^{\frac{1}{p}} < \infty,$$

where $\boldsymbol{x} := (x_1, x_2, x_3), \in \mathbb{R}^3$ and $d\boldsymbol{x} := dx_1 dx_2 dx_3$ stands for the usual Lebesgue measure on $\mathbb{R}^3$. For $p = \infty$, $L^\infty(\mathbb{R}^3, \mathbb{O})$ is the collection of essentially bounded measurable functions with the norm $\|f\|_\infty = ess\ sup_{\boldsymbol{x}\in\mathbb{R}^3}|f(\boldsymbol{x})|$. The next technical lemma yields the equivalence between the membership of an octonion function f in the octonion L^1-space (respectively L^2-space) and the membership in

the quaternion L^1-space (respectively L^2-space) of the components of f in its quaternion expression.
Expressing an octonion signal with the quaternion form as in (1), we obtain

$$f = g + h \odot e_4, \tag{12}$$

where g and h are two quaternion signals, then we get the equivalences:

Lemma 4.

(i) $f \in L^1(\mathbb{R}^3, \mathbb{O})$ *if and only if* $g, h \in L^1(\mathbb{R}^3, \mathbb{H})$,
(ii) $f \in L^2(\mathbb{R}^3, \mathbb{O})$ *if and only if* $g, h \in L^2(\mathbb{R}^3, \mathbb{H})$.

Proof. (i) Suppose that $f \in L^1(\mathbb{R}^3, \mathbb{O})$. Regarding the fact that $|g| \leq |f|$ and $|h| \leq |f|$, we obtain that $g, h \in L^1(\mathbb{R}^3, \mathbb{O})$. The reciprocal implication holds true since $|f| = |g + h \odot e_4| \leq |g| + |h|$.
(ii) Both implications are a consequence of the identity $|f|^2 = |g|^2 + |h|^2$.

2.2 Clifford Algebra $C\ell_{0,3}$ and Its Clifford-Fourier Transform

The real Clifford algebra $C\ell_{0,3}$ is Clifford's geometric algebra over $\mathbb{R}^3$, i.e. $C\ell_{0,3}$ is 8D linear space with basis:

$$\{1, e_1, e_2, e_3, e_{12}, e_{13}, e_{23}, i_3\}, \tag{13}$$

where 1 is the unit element, and we used the conventional index notation $e_{12} = e_1 \circ e_2, e_{13} = e_1 \circ e_3, e_{23} = e_2 \circ e_3, i_3 = e_1 \circ e_2 \circ e_3$, here the symbol $\circ$ stands for the Clifford multiplication.
The associative geometric multiplication of the basis vectors obeys to the laws

$$\begin{cases} e_k \circ e_k = e_k^2 = -1 & k \in \{1,2,3\}, \\ e_k \circ e_l = -e_l \circ e_k & k \neq l,\ k,l \in \{1,2,3\} \end{cases}$$

Table 2. $C\ell_{0,3}$ Multiplication Table

	Basis elements							
$\circ$	1	e_1	e_2	e_3	e_{12}	e_{13}	e_{23}	i_3
1	1	e_1	e_2	e_3	e_{12}	e_{13}	e_{23}	i_3
e_1	e_1	-1	e_{12}	e_{13}	$-e_2$	$-e_3$	i_3	$-e_{23}$
e_2	e_2	$-e_{12}$	-1	e_{23}	e_1	$-i_3$	$-e_3$	e_{13}
e_3	e_3	$-e_{13}$	$-e_{23}$	-1	i_3	e_1	e_2	$-e_{12}$
e_{12}	e_{12}	e_2	$-e_1$	i_3	-1	e_{23}	$-e_{13}$	$-e_3$
e_{13}	e_{13}	e_3	$-i_3$	$-e_1$	$-e_{23}$	-1	e_{12}	e_2
e_{23}	e_{23}	i_3	e_3	$-e_2$	e_{13}	$-e_{12}$	-1	$-e_1$
i_3	i_3	$-e_{23}$	e_{13}	$-e_{12}$	$-e_3$	e_2	$-e_1$	1

Clifford algebra $C\ell_{0,3}$ multiplication is given in Table 2, from which observe that $e_1, e_2, e_3, e_{12}, e_{13}, e_{23}$ are square roots of -1.
We recall that the 3D Clifford–Fourier transform is defined by for $f \in L^1(\mathbb{R}^3, C\ell_{0,3})$, by (see [2]):

$$\mathcal{F}_{3D}[f](w_1, w_2, w_3) = \int_{\mathbb{R}^3} f(x_1, x_2, x_3) \circ e^{-e_1 2\pi w_1 x_1} \circ e^{-e_2 2\pi w_2 x_2} \circ e^{-e_3 2\pi w_3 x_3}\, dx, \tag{14}$$

Remark 1. Although the octonion algebra and the $C\ell_{0,3}$-algebra are both of dimension 8 and non-commutative, they are clearly not identifiable with each other because the latter is associative and the other is not. They do, however, share the fact that one can dive into both of them the Euclidean space $\mathbb{R}^3$, the algebra of reals $\mathbb{R}$ (isomorphic to the Clifford algebra $C\ell_{0,0}$), the algebra of complex numbers $\mathbb{C}$ (isomorphic to the Clifford algebra $C\ell_{0,1}$), and the algebra of quaternions $\mathbb{H}$ (isomorphic to the Clifford algebra $C\ell_{0,2}$). Therefore, the multiplications $\odot$ and $\circ$ match on the quaternions, and in particular, on the complex and real numbers where the two symbols will be omitted. Furthermore, by using the formula (1), one can switch from an algebra of octonions problem to the Clifford algebra $C\ell_{0,3}$ restricted to the algebra of quaternions on $\mathbb{R}^3$, which is associative and for which the results of the Fourier–Clifford analysis are established, as it is an instance of the general real Clifford algebra $C\ell_{0,n}$.

3 Octonion Fourier Transform

In this section, we recall the definition of the octonion Fourier transform (OFT), outline some of its important results used in the sequel, and then add some new results. More details have been discussed in [10–12].
In what follows, we often use the shorthand $\boldsymbol{x} = (x_1, x_2, x_3)$, $\boldsymbol{w} = (w_1, w_2, w_3) \in \mathbb{R}^3$ and the $\mathbb{R}^3$-Lebesgue measure as $d\boldsymbol{x} = dx_1 dx_2 dx_3$.

Definition 1. *If $f \in L^1(\mathbb{R}^3, \mathbb{O})$, then the OFT of f is defined as follows:*

$$\mathcal{F}_{\mathbb{O}}[f](\boldsymbol{w}) = \int_{\mathbb{R}^3} f(\boldsymbol{x}) \odot e^{-e_1 2\pi w_1 x_1} \odot e^{-e_2 2\pi w_2 x_2} \odot e^{-e_4 2\pi w_3 x_3}\, d\boldsymbol{x}, \tag{15}$$

The OFT shares many properties with the classical (complex) and quaternion Fourier transforms (for more details on the properties of the OFT, see [3,4]).

Proposition 1. *The octonion Fourier transform $\mathcal{F}_{\mathbb{O}}$, enjoys the following properties:*
P(1) $\mathcal{F}_{\mathbb{O}}$ is a $\mathbb{R}$-linear:

$$\mathcal{F}_{\mathbb{O}}[\alpha f_1 + \beta f_2] = \alpha \mathcal{F}_{\mathbb{O}}[f_1] + \beta \mathcal{F}_{\mathbb{O}}[f_2], \quad \alpha, \beta \in \mathbb{R}.$$

P(2) Shift property: For $\alpha, \beta, \gamma \in \mathbb{R}$

$$\mathcal{F}_{\mathbb{O}}[f((x_1 - \alpha, x_2, x_3)](\boldsymbol{w}) = \cos(2\pi x_1 \alpha) U(\boldsymbol{w}) - \sin(2\pi x_1 \alpha) \mathcal{F}_{\mathbb{O}}[f](x_1, -x_2, -x_3) \odot e_1,$$
$$\mathcal{F}_{\mathbb{O}}[f((x_1, x_2 - \beta, x_3)](\boldsymbol{w}) = \cos(2\pi x_2 \beta) \mathcal{F}_{\mathbb{O}}[f](\boldsymbol{w}) - \sin(2\pi x_2 \beta) \mathcal{F}_{\mathbb{O}}[f](w_1, w_2, -w_3) \odot e_2,$$
$$\mathcal{F}_{\mathbb{O}}[f((x_1, x_2, x_3 - \gamma)](\boldsymbol{w}) = \cos(2\pi x_3 \gamma) \mathcal{F}_{\mathbb{O}}[f](\boldsymbol{w}) - \sin(2\pi x_3 \gamma) \mathcal{F}_{\mathbb{O}}[f](\boldsymbol{w}) \odot e_4.$$

P(3) Scaling property: For $\alpha, \beta, \gamma \in \mathbb{R}\backslash\{0\}$

$$\mathcal{F}_{\mathbb{O}}\left[f\left(\frac{x_1}{\alpha}, \frac{x_2}{\beta}, \frac{x_3}{\gamma}\right)\right](\boldsymbol{w}) = |\alpha\beta\gamma|\mathcal{F}_{\mathbb{O}}[f](\alpha w_1, \beta w_2, \gamma w_3).$$

P(4) The Riemann-Lebesgue theorem $\lim_{|\boldsymbol{w}|\to\infty} \mathcal{F}_{\mathbb{O}}[f](\boldsymbol{w}) = 0.$

The OFT is not an $\mathbb{O}$-linear operation (see [12, Lemma 3.7]); thus $P(1)$ does not hold for any $\alpha, \beta \in \mathbb{O}$, because the multiplication of octonions is not associative. The following theorem demonstrates that the OFT is invertible (see [4]).

Theorem 3 (Inversion formula). *For* $f \in L^1(\mathbb{R}^3, \mathbb{O})$, *such that* $\mathcal{F}_{\mathbb{O}}[f] \in L^1(\mathbb{R}^3, \mathbb{O})$, *the inverse of the OFT can be computed as follows:*

$$f(\boldsymbol{x}) = \int_{\mathbb{R}^3} \mathcal{F}_{\mathbb{O}}[f](\boldsymbol{w}) \odot e^{e_4 2\pi w_3 x_3} \odot e^{e_2 2\pi w_2 x_2} \odot e^{e_1 2\pi w_1 x_1}\, d\boldsymbol{w}, \tag{16}$$

The next example indicates that the OFT of a Gaussian octonion function is another Gaussian octonion function.

Example 1 (OFT of a Gaussian octonion function). Consider a Gaussian octonion function f given by $f(\boldsymbol{x}) = oe^{-\pi|\boldsymbol{x}|^2}$, where $o = a + b \odot e_4$ is a constant quaternion. Then

$$\mathcal{F}_{\mathbb{O}}[f](\boldsymbol{w}) = f(\boldsymbol{w}). \tag{17}$$

The following lemma shows that the OFT retains the energy of octonion-valued signals. [4, Theorem 18]:

Lemma 5 (Parseval theorem). *For* $f \in L^2(\mathbb{R}^3, \mathbb{O})$, *one has*

$$\left\|\mathcal{F}_{\mathbb{O}}[f]\right\|_2 = \|f\|_2. \tag{18}$$

Remark 2. $L^1(\mathbb{R}^3, \mathbb{O}) \cap L^2(\mathbb{R}^3, \mathbb{O})$, like in the classical case, is dense in $L^2(\mathbb{R}^3, \mathbb{O})$. Thus, standard reasoning on density leads us to extend the definition of the OFT of $f \in L^1(\mathbb{R}^3, \mathbb{O}) \cap L^2(\mathbb{R}^3, \mathbb{O})$ in a unique way to the whole of $L^2(\mathbb{R}^3, \mathbb{O})$. Hence, we consider that the definition of the OFT is an operator of $L^2(\mathbb{R}^3, \mathbb{O})$ into $L^2(\mathbb{R}^3, \mathbb{O})$.

In the following, the even and odd parts are denoted by f_e and the f_o, receptively, in the third variable x_3, of f, given by the following:

$$f_e = \frac{1}{2}\left(f(x_1, x_2, x_3) + f(x_1, x_2, -x_3)\right), \text{and } f_o = \frac{1}{2}\left(f(x_1, x_2, x_3) - f(x_1, x_2, -x_3)\right). \tag{19}$$

We prove the following formula, which is based on the norm of an octonion signal f in terms of the norms of its odd and even parts in its quaternion decomposition, using long but simple calculations.

Lemma 6.

$$\frac{1}{2}\left(|f|^2 + |f(x_1, x_2, -x_3)|^2\right) = |g_e|^2 + |g_o|^2 + |h_e|^2 + |h_o|^2. \tag{20}$$

The OFT and 3D-Clifford-Fourier transform have the following relationship:

Lemma 7.

$$\mathcal{F}_{\mathbb{O}}[f] = \mathcal{F}_{3D}[f_+] + \mathcal{F}_{3D}[f_-] \odot e_4. \tag{21}$$

where

$$f_+ = g_e + h_o \odot e_3, \;\; f_-(\boldsymbol{x}) = h_e(-x_1, -x_2, x_3) - g_o(-x_1, -x_2, x_3) \odot e_3 \tag{22}$$

and g, h are the quaternion parts in the decomposition (12) of f.
Lemma 7 can be proved using Lemma 3 and the Euler formula for the octonions (7).

Remark 3. The lemma is significant insofar as it makes it possible to note that the computations of the OFT of an octonion-valued signal are reduced to the computations of the 3D Clifford-Fourier transform of a quaternion-valued signals.

Given that the 3D Clifford transforms of f_+ and f_- lie in the quaternion algebra and considering the two formulas (21) and (10), we have the following lemma:

Lemma 8. *For* $f \in L^1(\mathbb{R}^3, \mathbb{O})$, *we have*

$$\left|\mathcal{F}_{\mathbb{O}}[f]\right|^2 = \left|\mathcal{F}_{3D}[f_+]\right|^2 + \left|\mathcal{F}_{3D}[f_-]\right|^2. \tag{23}$$

We define a new module of $\mathcal{F}_{\mathbb{O}}[f]$ as follows :

$$|\mathcal{F}_{\mathbb{O}}[f]|_{\mathcal{O}} := \sqrt{\sum_{m=0}^{m=7} |\mathcal{F}_{\mathbb{O}}[f_m]|^2}. \tag{24}$$

Furthermore, we define a new L^2-norm of $\mathcal{F}_{\mathbb{O}}[f]$ as follows

$$\|\mathcal{F}_{\mathbb{O}}[f]\|_{2,\mathcal{O}} := \sqrt{\int_{\mathbb{R}^3} \left|\mathcal{F}_{\mathbb{O}}[f](\boldsymbol{w})\right|_{\mathcal{O}}^2 d\boldsymbol{w}}. \tag{25}$$

It is interesting to observe that $|\mathcal{F}_{\mathbb{O}}[f]|_{\mathcal{O}}$ is not equivalent to $|\mathcal{F}_{\mathbb{O}}[f]|$ unless f is real valued.

4 Uncertainty Relations for the Octonion Fourier Transform

The classical UP is a fundamentally accurate result of signal processing and illustrates the precision with which a signal can be measured in space and in its spectral (frequency) domain. UPs have recently been studied for hypercomplex Fourier signals, including quaternion-valued signals, space-time-valued signals and Clifford-Fourier signals. See, for example, [5,8].

Inspired by the generalization of Beurling's theorem in terms of the quaternion algebra [5], we investigate in this section Beurling's UP for the OFT.

4.1 Beurling's up

The following useful proposition is an extension of Beurling's UP for the quaternion Fourier transform [5, Thm. 4.2] on $\mathbb{R}^2$ to the 3–D quaternion Fourier transform signals.

Proposition 2. *Let $f \in L^2(\mathbb{R}^3, \mathbb{R})$ and suppose that*

$$\int_{\mathbb{R}^3}\int_{\mathbb{R}^3} |f(\boldsymbol{x})||\mathcal{F}_{3D}[f](\boldsymbol{w})|e^{2\pi|\boldsymbol{x}||\boldsymbol{w}|}d\boldsymbol{x}d\boldsymbol{w} < \infty. \tag{26}$$

Then $f = 0$ almost everywhere.

Now, according to Proposition 2, we prove an analogue of Theorem 1 for the OFT.

Theorem 4 (Beurling's UP). *Suppose $f \in L^2(\mathbb{R}^3, \mathbb{O})$ with*

$$\int_{\mathbb{R}^3}\int_{\mathbb{R}^3} |f(\boldsymbol{x})||\mathcal{F}_{\mathbb{O}}[f](\boldsymbol{w})|_{\mathcal{O}}e^{2\pi|\boldsymbol{x}||\boldsymbol{w}|}d\boldsymbol{x}d\boldsymbol{w} < \infty. \tag{27}$$

Then $f = 0$ almost everywhere.

Proof. We can assume without loss of generality that $f \in L^2(\mathbb{R}^3, \mathbb{R})$. Indeed, if we suppose that the result is proven for a real-valued function $f_m \in L^2(\mathbb{R}^3, \mathbb{R})$, which is a component of f given by the form (11), then the assumption (27) implies that

$$\int_{\mathbb{R}^3}\int_{\mathbb{R}^3} |f_m(\boldsymbol{x})||\mathcal{F}_{\mathbb{O}}[f_m](\boldsymbol{w})|e^{2\pi|\boldsymbol{x}||\boldsymbol{w}|}d\boldsymbol{x}d\boldsymbol{w} < \infty,$$

hence we will have f_m is 0 almost everywhere. and therefore so is f.

Now, let $f \in L^2(\mathbb{R}^3, \mathbb{R})$, and $f = g + h \odot e_4$ be the quaternion form of f given by (12), and let $f_+; f_-$ be the functions given by (22), then we have $f = g, h = 0$, and consequently $f_+ = f_e$ and $f_- = -f_o(-x_1, -x_2, x_3) \odot e_3$.
By Lemma 6 we get that f_+ and f_- are both in $L^2(\mathbb{R}^3, \mathbb{R})$.

Moreover, by noticing from Lemma 8 that $|\mathcal{F}_{3D}[f_\pm]| \leq |\mathcal{F}_{\mathbb{O}}[f]|$, the assumption (27) yields that

$$\int_{\mathbb{R}^3}\int_{\mathbb{R}^3} |f_\pm(\boldsymbol{x})||\mathcal{F}_{3D}[f_\pm](\boldsymbol{w})|e^{2\pi|\boldsymbol{x}||\boldsymbol{w}|}d\boldsymbol{x}d\boldsymbol{w} < \infty,$$

which is equivalent to

$$\int_{\mathbb{R}^3}\int_{\mathbb{R}^3} |f_e(\boldsymbol{x})||\mathcal{F}_{3D}[f_e(\boldsymbol{w})|e^{2\pi|\boldsymbol{x}||\boldsymbol{w}|}d\boldsymbol{x}d\boldsymbol{w} < \infty,$$

and

$$\int_{\mathbb{R}^3}\int_{\mathbb{R}^3} |-f_o(-x_1,-x_2,x_3)\odot e_3|\mathcal{F}_{3D}[-f_o(-x_1,-x_2,x_3)\odot e_3](\boldsymbol{w})|e^{2\pi|\boldsymbol{x}||\boldsymbol{w}|}d\boldsymbol{x}d\boldsymbol{w} < \infty.$$

As $|-f_o(-x_1,-x_2,x_3)\odot e_3| = |f_o(-x_1,-x_2,x_3)|$, we can, easily, prove the last inequality is alternatively

$$\int_{\mathbb{R}^3}\int_{\mathbb{R}^3} |f_o||\mathcal{F}_{3D}[f_o](\boldsymbol{w})|e^{2\pi|\boldsymbol{x}||\boldsymbol{w}|}d\boldsymbol{x}d\boldsymbol{w} < \infty.$$

However, based on Proposition 2, we get $f_e = f_o = 0$, therefore $f = f_e + f_o = 0$.

Corollary 1 (Hardy's UP).
Let $f \in L^2(\mathbb{R}^3, \mathbb{O})$. Suppose for some $\alpha, \beta > 0, f$ satisfies

$$|f(\boldsymbol{x})| \leq Ce^{-\pi\alpha|x|^2}, \text{ and } |\mathcal{F}_{\mathbb{O}}[f](\boldsymbol{w})|_{\mathcal{O}} \leq C'e^{-\pi\beta|\boldsymbol{w}|^2},$$

where C and C' are positive constants. If moreover

$$\alpha\beta > 1$$

then $f = 0$ almost everywhere.

Corollary 2 (Gelfand–Shilov's UP).
Let $f \in L^2(\mathbb{R}^3, \mathbb{O})$, and assume that

$$\int_{\mathbb{R}^3} |f(\boldsymbol{x})|e^{2\pi\frac{\alpha^p}{p}|x|^p}d\boldsymbol{x} < \infty, \text{ and } \int_{\mathbb{R}^3} |\mathcal{F}_{\mathbb{O}}[f](\boldsymbol{w})|_{\mathcal{O}}e^{2\pi\frac{\beta^q}{q}|w|^q}d\boldsymbol{w} < \infty, \tag{28}$$

for some $\alpha, \beta > 0$, $1 < p, q < \infty$ with $\dfrac{1}{p} + \dfrac{1}{q} = 1$. Then $f = 0$ almost everywhere whenever $(p, q) \neq (2, 2)$ or $\alpha\beta > 1$.

Corollary 3 (Cowling–Price's UP).
Let $f \in L^2(\mathbb{R}^3, \mathbb{O})$, and assume that

$$\int_{\mathbb{R}^3} \left(|f(\boldsymbol{x})|e^{\pi\alpha|x|^2}\right)^p d\boldsymbol{x} < \infty, \int_{\mathbb{R}^3} \left(|\mathcal{F}_{\mathbb{O}}[f](\boldsymbol{w})|_{\mathcal{O}}e^{\pi\beta|y|^2}\right)^q dy < \infty,$$

with $1 < p, q < \infty$, $\dfrac{1}{p} + \dfrac{1}{q} = 1$. If $\alpha\beta > 1$, then $f = 0$ almost everywhere.

5 Conclusion

In the present paper, we first proved an analogue of Beurling's theorem in the framework of octonions based on the quaternion form of octonions, the version of Beurling's uncertainty principle related to the quaternion algebra, and the relation between the Fourier transform of octonions on $\mathbb{R}^2$ and the 3D Clifford-Fourier transform. We then derived other variants of the uncertainty principle.

Acknowledgements. The authors extend their appreciation to the Deanship of Scientific Research at King Khalid University, Saudi Arabia, for funding this work through the research groups program under grant R.G.P.1/207/43. They further thank E. Hitzer, as well as the anonymous referees and the organizers of CGI 2022.

References

1. Bagchi, S.C., Ray, S.K.: Uncertainty principles like Hardy's theorem on some Lie groups. J. Austral. Math. Soc. Ser. A **65**, 289–302 (1999)
2. Brackx, F., Hitzer, E., Sangwine, S.: History of quaternion and Clifford-Fourier transforms. In: Hitzer, E., Sangwine, S.J. (eds.) Quaternion and Clifford-Fourier Transforms and Wavelets. Trends in Mathematics (TIM), vol. 27, pp. xi-xxvii. Birkhauser, Basel (2013)
3. Błaszczyk, Ł.: Octonion spectrum of 3D octonion-valued signals-properties and possible applications. In: 26th European Signal Processing Conference (EUSIPCO), pp. 509–513 (2018). https://doi.org/10.23919/EUSIPCO.2018.8553228
4. Błaszczyk, Ł: A generalization of the octonion Fourier transform to 3-D octonion-valued signals: properties and possible applications to 3-D LTI partial differential systems. Multidimension. Syst. Signal Process. **31**(4), 1227–1257 (2020). https://doi.org/10.1007/s11045-020-00706-3
5. El Haoui, Y., Fahlaoui, S.: Beurling's theorem for the quaternion Fourier transform. J. Pseudo-Differ. Operators Appl. **11**(1), 187–199 (2019). https://doi.org/10.1007/s11868-019-00281-7
6. Folland, G.B., Sitaram, A.: The uncertainty principle: a mathematical survey. J. Fourier Anal. Appl. **3**, 207–238 (1997)
7. Hardy, G.H.: A theorem concerning Fourier transform. J. London Math. Soc. **8**, 227–231 (1933). https://doi.org/10.1112/jlms/s1-8.3.227
8. Hitzer, E.: Special affine Fourier transform for space-time algebra signals. In: Magnenat-Thalmann, N., et al. (eds.) CGI 2021. LNCS, vol. 13002, pp. 658–669. Springer, Cham (2021). https://doi.org/10.1007/978-3-030-89029-2_49
9. Hörmander, L.: A uniqueness theorem of Beurling for Fourier transform pairs. Ark. För Math. **2**, 237–240 (1991)
10. Kauhanen, J., Orelma, H.: Cauchy–Riemann operators in octonionic analysis. Adv. Appl. Clifford Algebras **28**(1), 1–14 (2018). https://doi.org/10.1007/s00006-018-0826-2
11. Lian, P.: The octonion Fourier transform: uncertainty relations and convolution. Signal Process. **164**, 295–300 (2019). https://doi.org/10.1016/j.sigpro.2019.06.015
12. Li, Y., Ren, G.: Real Paley-Wiener theorem for the octonion Fourier transform. Math. Meth. Appl. Sci., 1–16 (2021). https://doi.org/10.1002/mma.7513

Embedding of Octonion Fourier Transform in Geometric Algebra of $\mathbb{R}^3$ and Polar Representations of Octonion Analytic Signals

Eckhard Hitzer(✉)

International Christian University, Mitaka, Tokyo 181-8585, Japan
hitzer@icu.ac.jp
https://geometricalgebrajp.wordpress.com/

Abstract. We show how the octonion Fourier transform can be embedded and studied in Clifford geometric algebra of three-dimensional Euclidean space $Cl(3,0)$. We apply a new form of dimensionally minimal embedding of octonions in geometric algebra, that expresses octonion multiplication non-associativity with a sum of up to four (individually associative) geometric algebra product terms. This approach leads to new polar representations of octonion analytic signals.

Keywords: Clifford geometric algebra · octonions · Fourier transform · analytic signal · polar representation

1 Introduction

Hypercomplex Fourier transforms experienced rapid development during the last 30 years. A historical overview of this field can be found in [3], a variety of approaches is included in [7], and a recent comprehensive textbook is [8]. For a recent survey of signal and image processing in Clifford geometric algebra, see Sect. 6 of [10]. In Definition 9 of [4] a Clifford algebra based hypercomplex Fourier transform producing a multidimensional analytic signal was defined. In the book [5] this approach is applied for the non-associative and non-commutative hypercomplex algebra of octonions. Apart from its non-associativity, octonions have many outstanding algebraic properties (e.g. the highest dimensional normed division algebra). It is therefore of great interest for us in this work to use a recently invented minimal embedding [11,12] of octonions in the Clifford geometric algebra of three-dimensional space $Cl(3,0)$ and consequently embed the octonion Fourier transformation (OFT) in $Cl(3,0)$. This embedding allows to break down non-associative octonion multiplication into sums of associative geometric products, and therefore to easily apply existing geometric algebra computing software

Soli Deo Gloria. Dedicated to the 44,821 Europeans who died due to Covid-19 vaccine adverse drug reactions, as of 21 May 2022 [17].

E. Hitzer et al. (Eds.): ENGAGE 2022, LNCS 13862, pp. 123–134, 2023.
https://doi.org/10.1007/978-3-031-30923-6_10

[1,2,15]. And it allows to establish new polar representations for octonion analytic signals.

We first review in Sect. 2 the properties of octonions [13] and in Sect. 3 the new embedding of octonions in Clifford geometric algebra $Cl(3,0)$. Then we present in Sect. 4 the OFT of [5], as well as octonion analytic signals, and in Sect. 5 embed the OFT in $Cl(3,0)$. Finally, in Sect. 6 we utilize the polar decomposition of [9,16] for complex biquaternions and multivectors in $Cl(3,0)$ to introduce new polar representations for octonion analytic signals.

2 Octonions

Here we first briefly summarize important octonion algebra properties (see [13], pp. 300–302, and [11]), assuming $a, b, c, x, y \in \mathbb{O}$.

- Octonions $\mathbb{O}$ form an eight-dimensional bilinear algebra over the reals $\mathbb{R}$ with basis $\{1, \mathbf{e}_1, \mathbf{e}_2, \mathbf{e}_3, \mathbf{e}_4, \mathbf{e}_5, \mathbf{e}_6, \mathbf{e}_7\}$.
- The multiplication table[1] is given by $(1 \leq i, j \leq 7)$
$$\mathbf{e}_i \star \mathbf{e}_i = -1, \quad \mathbf{e}_i \star \mathbf{e}_j = -\mathbf{e}_j \star \mathbf{e}_i \text{ for } i \neq j, \qquad \mathbf{e}_i \star \mathbf{e}_{i+1} = \mathbf{e}_{i+3}, \tag{1}$$
where $(i, i+1, i+3)$ can be permuted cyclically and translated modulo 7.
- Via the Cayley-Dickson doubling process, octonions can directly be defined from pairs of quaternions $p_1, p_2, q_1, q_2 \in \mathbb{H}$ (note the order of factors, qc(...) is quaternion conjugation):
$$(p_1, q_1) \star (p_2, q_2) = \big(p_1 p_2 - \mathrm{qc}(q_2) q_1,\ q_2 p_1 + q_1 \mathrm{qc}(p_2)\big). \tag{2}$$
- $\mathbb{O}$ has no zero divisors, i.e., $ab = 0$ implies $a = 0$ or $b = 0$.
- $\mathbb{O}$ is a division algebra, i.e., $ax = b$ and $ya = b$ have unique solutions x, y for non-zero a.
- $\mathbb{O}$ admits unique inverses.
- $\mathbb{O}$ is non-associative, i.e., in general $a(bc) \neq (ab)c$.
- $\mathbb{O}$ is alternative, i.e., $a(ab) = a^2 b$ and $(ab)b = ab^2$.
- $\mathbb{O}$ is one of only four alternative division algebras over $\mathbb{R}$: $\mathbb{R}, \mathbb{C}, \mathbb{H}, \mathbb{O}$.
- $\mathbb{O}$ is flexible, i.e., $a(ba) = (ab)a$.
- $\mathbb{O}$ has a (positive-definite quadratic form) norm $\|\ldots\| : \mathbb{O} \to \mathbb{R}$, the norm is preserved (i.e. admits composition), such that $\|ab\| = \|a\|\|b\|$.
- $\mathbb{O}$ is one of only four unital norm-preserving division algebras over $\mathbb{R}$: $\mathbb{R}, \mathbb{C}, \mathbb{H}, \mathbb{O}$.
- $\mathbb{O}$ is essential for treating *triality*, an automorphism of the universal covering spin group Spin(8) of the rotation group SO(8) or $\mathbb{R}^8$. Triality is not an inner automorphism, nor an orthogonal matrix similarity, nor a linear transformation $Cl(8,0) \to Cl(8,0)$, nor a linear automorphism of SO(8). Triality permutes three elements in the center of $Cl(8,0)$, namely $\{-1, e_{12345678}, -e_{12345678}\}$, with basis vectors e_i, $(1 \leq i \leq 8)$, of $\mathbb{R}^8$. Triality is a restriction of a polynomial mapping $Cl(8,0) \to Cl(8,0)$ of degree two.

Furthermore, like for complex numbers, quaternions and biquaternions, there is a *polar decomposition* for octonions [16].

[1] This depends obviously on deliberate ordering and sign choices for the basis elements

3 Embedding of Octonions in Clifford Geometric Algebra of Three-Dimensional Euclidean Space

For readers not familiar with Clifford geometric algebra we refer to the excellent textbook [13], and to the tutorial introduction [6]. The current section summarizes the results needed from [11].

The Clifford geometric algebra $Cl(3,0)$ of Euclidean space $\mathbb{R}^3$ has eight basis elements

$$\{1, \sigma_1, \sigma_2, \sigma_3, I\sigma_1 = \sigma_{23}, I\sigma_2 = \sigma_{31}, I\sigma_3 = \sigma_{12}, I = \sigma_{123}\}, \tag{3}$$

where $\{\sigma_1, \sigma_2, \sigma_3\}$ forms an orthonormal vector basis of $\mathbb{R}^3$. We can construct in $Cl(3,0)$ an octonionic product [11], after splitting it in its even subalgebra $Cl^+(3,0)$ with basis

$$\{1, \sigma_{23}, \sigma_{31}, \sigma_{12}\}, \tag{4}$$

and the set $Cl^-(3,0)$ of odd grade (w.r.t. grades in $Cl(3,0)$) elements

$$\{\sigma_1, \sigma_2, \sigma_3, I = \sigma_{123}\}. \tag{5}$$

We will use the Clifford conjugation[2] (indicated by an overbar $\overline{M}$), i.e. the composition of (main) grade involution ($\widehat{M}$) and reversion ($\widetilde{M}$), which preserves grades zero and three, but changes the signs of grades one and two in $Cl(3,0)$. A realization of the octonionic product of M, N in $Cl(3,0)$ is given by four geometric algebra product terms

$$\begin{aligned} M &= M_+ + M_-, \quad N = N_+ + N_-, \\ M \star N &= M_+N_+ + \overline{N_-}M_- + N_-M_+ + M_-\overline{N_+}, \end{aligned} \tag{6}$$

with even grade parts $M_+, N_+ \in Cl^+(3,0)$ and odd grade parts $M_-, N_- \in Cl^-(3,0)$. The multiplication table is Table 1, with octonionic product illustration in Fano plane diagram form in Fig. 1.

The octonion conjugate (anti-involution) in $Cl(3,0)$ is given by

$$M^* = \widetilde{M}_+ - M_- = \overline{M}_+ - M_-, \quad (M \star N)^* = N^* \star M^*. \tag{7}$$

Computing the octonion norm yields (including norm-preservation):

$$\|M\| = M \star M^* = \langle M\widetilde{M}\rangle = M * \widetilde{M} = \sum_{i=1}^{8} M_i^2, \qquad \|M \star N\| = \|M\|\|N\|. \tag{8}$$

where $M_i \in \mathbb{R}, 1 \leq i \leq 8$, are the coefficients of M in the $Cl(3,0)$ basis (3).

The above reviewed embedding is very flexible. It even allows to reversely embed Clifford geometric algebra $Cl(3,0)$ in octonions by defining the geometric product in terms of the octonionic product (see [11], Sect. 3.3 for details):

$$\begin{aligned} &M_+N_+ \overset{(6)}{=} M_+ \star N_+, \quad M_-N_- \overset{(6)}{=} N_- \star \overline{M}_-, \\ &M_-N_+ \overset{(6)}{=} N_+ \star M_-, \quad M_+N_- = -(N_- \star I) \star (M_+ \star I). \end{aligned} \tag{9}$$

[2] Note that by construction $\overline{M_\pm} = (\overline{M})_\pm$.

Table 1. Multiplication table for octonion embedding in $Cl(3,0)$. The upper left 4×4-block corresponds to M_+N_+, the upper right 4×4-block to N_-M_+, the lower left 4×4-block to $M_-\overline{N}_+$, and the lower right 4×4-block to $\overline{M}_-N_-$ of (6).

Left factors	Right factors							
	1	$I\sigma_1$	$I\sigma_2$	$I\sigma_3$	σ_1	σ_2	σ_3	I
1	1	$I\sigma_1$	$I\sigma_2$	$I\sigma_3$	σ_1	σ_2	σ_3	I
$I\sigma_1$	$I\sigma_1$	-1	$-I\sigma_3$	$I\sigma_2$	I	σ_3	$-\sigma_2$	$-\sigma_1$
$I\sigma_2$	$I\sigma_2$	$I\sigma_3$	-1	$-I\sigma_1$	$-\sigma_3$	I	σ_1	$-\sigma_2$
$I\sigma_3$	$I\sigma_3$	$-I\sigma_2$	$I\sigma_1$	-1	σ_2	$-\sigma_1$	I	$-\sigma_3$
σ_1	σ_1	$-I$	σ_3	$-\sigma_2$	-1	$I\sigma_3$	$-I\sigma_2$	$I\sigma_1$
σ_2	σ_2	$-\sigma_3$	$-I$	σ_1	$-I\sigma_3$	-1	$I\sigma_1$	$I\sigma_2$
σ_3	σ_3	σ_2	$-\sigma_1$	$-I$	$I\sigma_2$	$-I\sigma_1$	-1	$I\sigma_3$
I	I	σ_1	σ_2	σ_3	$-I\sigma_1$	$-I\sigma_2$	$-I\sigma_3$	-1

4 Octonion Fourier Transform

From now on, if no brackets are given, the order of multiplication is assumed to be from left to right. According to Sect. 4.2.1 of [5], the OFT of an integrable real signal $f \in L^1(\mathbb{R}^3,\mathbb{R})$ can be defined as

$$\mathcal{F}\{f\}(\mathbf{u}) = \int_{\mathbb{R}^3} f(\mathbf{x})e^{-\mathbf{e}_1 2\pi u_1 x_1}e^{-\mathbf{e}_2 2\pi u_2 x_2}e^{-\mathbf{e}_4 2\pi u_3 x_3}d^3x, \tag{10}$$

with three-dimensional position- and frequency vectors, and volume element

$$\mathbf{x} = (x_1,x_2,x_3) \in \mathbb{R}^3, \quad \mathbf{u} = (u_1,u_2,u_3) \in \mathbb{R}^3, \quad d^3x = dx_1dx_2dx_3, \tag{11}$$

respectively, and octonion units $\{\mathbf{e}_1,\mathbf{e}_2,\mathbf{e}_4\}$ in the exponents. As pointed out in [5], any triplet of octonion units could be used in the octonionic kernel of (10), as long as the three do not form a quaternionic subalgebra, by that reason, e.g., the triplet $\{\mathbf{e}_1,\mathbf{e}_2,\mathbf{e}_3\}$ is excluded, compare the multiplication table Table 2.3 and its Fano plane visualization Fig. 2.2 in [5]. In the latter the triplet $\{\mathbf{e}_1,\mathbf{e}_2,\mathbf{e}_3\}$ clearly lies on a straight line.

Given suitable integrability conditions, the inverse OFT can be computed as

$$f(\mathbf{x}) = \mathcal{F}^{-1}\{\mathcal{F}\{f\}\}(\mathbf{x}) = \int_{\mathbb{R}^3} \mathcal{F}\{f\}(\mathbf{u})e^{\mathbf{e}_4 2\pi u_3 x_3}e^{\mathbf{e}_2 2\pi u_2 x_2}e^{\mathbf{e}_1 2\pi u_1 x_1}d^3u,$$
$$d^3u = du_1du_2du_3. \tag{12}$$

Abbreviating $s_k = \sin(2\pi u_k x_k)$, $c_k = \cos(2\pi u_k x_k)$, $k = 1,2,3$, we can express the kernel of (10), using multiplication table Table 2.3 of [5], as

$$\begin{aligned} e^{-\mathbf{e}_1 2\pi u_1 x_1}e^{-\mathbf{e}_2 2\pi u_2 x_2}e^{-\mathbf{e}_4 2\pi u_3 x_3} &= (c_1 - s_1\mathbf{e}_1)(c_2 - s_2\mathbf{e}_2)(c_3 - s_3\mathbf{e}_4) \\ &= c_1c_2c_3 - s_1c_2c_3\mathbf{e}_1 - c_1s_2c_3\mathbf{e}_2 - c_1c_2s_3\mathbf{e}_4 \\ &\quad + s_1s_2c_3\mathbf{e}_3 + s_1c_2s_3\mathbf{e}_5 + c_1s_2s_3\mathbf{e}_6 - s_1s_2s_3\mathbf{e}_7. \end{aligned} \tag{13}$$

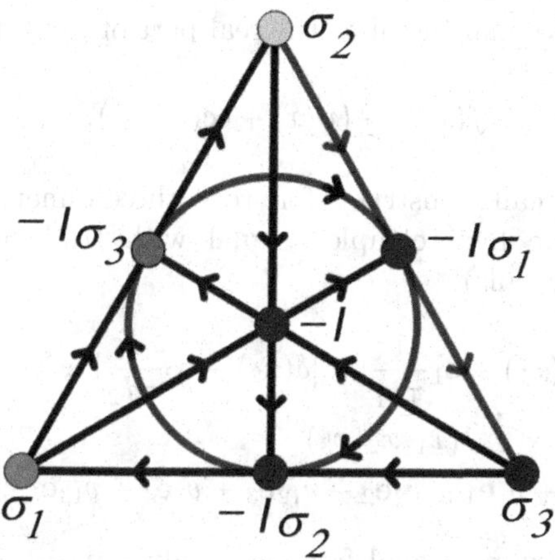

Fig. 1. Illustration of $Cl(3,0)$ basis elements under the octonionic product (6) in Table 1, see [11] for details.

The significance of this decomposition is, that therefore a real signal $f \in L^1(\mathbb{R}^3, \mathbb{R})$ is decomposed by the OFT (10) into eight spectral components of distinct even-odd symmetries: {eee,oee,eoe,eeo,ooe,oeo,eoo,ooo}, where e=even, o=odd. Following the multiplication table Table 2.3 of [5], and using the alternative octonion multiplication property of Sect. 2, we find the following conjugations $(i, j = 2, \ldots, 7)$

$$\alpha_i(\mathbf{e}_j) = \mathbf{e}_i\mathbf{e}_j\mathbf{e}_i = \begin{cases} \mathbf{e}_j, & i \neq j \\ -\mathbf{e}_j, & i = j \end{cases}. \tag{14}$$

This allows to express all $\mathcal{F}\{f\}(\pm u_1, \pm u_2, \pm u_3)$ in terms of $\mathcal{F}\{f\}(\mathbf{u})$ each time using four suitable α_i conjugations. For example,

$$\mathcal{F}\{f\}(-u_1, u_2, u_3) = \alpha_1(\alpha_3(\alpha_5(\alpha_7(\mathcal{F}\{f\}(\mathbf{u}))))). \tag{15}$$

As a consequence the OFT in all eight octants of the three-dimensional frequency space can be obtained from the OFT only applied to the first octant, where all three frequency components are positive (i.e. $\{u_1 \geq 0, u_2 \geq 0, u_3 \geq 0\}$).

4.1 Hypercomplex Analytic Signal

A real signal $f \in L^1(\mathbb{R}, \mathbb{R})$ can be extended to a complex analytic signal with *positive* frequency by multiplying its Fourier transform $\mathcal{F}_{\mathbb{R}}\{f\}(u)$ with $(1+\operatorname{sgn} u)$, $u \in \mathbb{R}$ being the frequency, and back transforming

$$\psi(x) = \mathcal{F}_{\mathbb{R}}^{-1}\{(1 + \operatorname{sgn} u)\mathcal{F}_{\mathbb{R}}\{f\}(u)\}(x), \tag{16}$$

equivalent to application of the Hilbert transform, where $\circledast$ means convolution,

$$H[f(x)] = (\frac{1}{\pi x}) \circledast f(x), \qquad \psi(x) = f(x) + iH[f(x)] = [\delta(x) + i\frac{1}{\pi x}] \circledast f(x). \tag{17}$$

We can recover the original signal as the real part of $\psi(x)$, i.e.,

$$f(x) = \frac{1}{2}(\psi(x) + \mathrm{cc}(\psi(x))). \tag{18}$$

Analogously, we can construct for real three-dimensional signals $f \in L^1(\mathbb{R}^3, \mathbb{R})$ an analytic hypercomplex signal with triple convolution by (see Sect. 5.2.3 of [5] for details)

$$\begin{aligned}\psi(x_1, x_2, x_3)_1 =&[\delta(x_1) + \mathbf{e}_1 \frac{1}{\pi x_1}] \times [\delta(x_2) + \mathbf{e}_2 \frac{1}{\pi x_2}] \times [\delta(x_3) + \mathbf{e}_4 \frac{1}{\pi x_3}] \\ &\circledast \circledast \circledast f(x_1, x_2, x_3) \\ =& f + v_1\mathbf{e}_1 + v_2\mathbf{e}_2 + v_{12}\mathbf{e}_3 + v_3\mathbf{e}_4 + v_{13}\mathbf{e}_5 + v_{23}\mathbf{e}_6 + v\mathbf{e}_7,\end{aligned} \tag{19}$$

which has only three-dimensional frequency values $\mathbf{u} = (u_1, u_2, u_3)$ in the first octant of frequency space, where all three frequency components are positive $(+++)$. The original signal $f \in L^1(\mathbb{R}^3, \mathbb{R})$ is the scalar real component of $\psi(x_1, x_2, x_3)$. The corresponding analytic signals $\psi(x_1, x_2, x_3)_k$, $k = 2, \ldots, 8$ in the other seven octants are obtained by simply changing the three plus signs in (19) to $(-++), (+-+), (--+), (++-), (-+-), (+--), (---)$, respectively. And we can recover the original signal simply by

$$f(x_1, x_2, x_3) = \frac{1}{8}\sum_{k=1}^{8} \psi(x_1, x_2, x_3)_k. \tag{20}$$

Instead of computing $\psi(x_1, x_2, x_3)_k$, $k = 2, \ldots, 8$, one by one, we can obviously also obtain them from $\psi(x_1, x_2, x_3)_1$ by applying to it compositions of octonionic conjugations (14) as, e.g., in (15). We note that [5], p. 167, states for $\psi(x_1, x_2, x_3)_1$ of (19): *The exact polar representation of this signal is unknown.*

This outline of the OFT and its corresponding analytic first octant frequency spectrum signal may suffice here to be able to somewhat appreciate its uniquely interesting properties, due to its octonionic kernel. For more details we refer to [5].

5 Embedding the OFT in Clifford Geometric Algebra of Three-Dimensional Euclidean Space

Now we reach the main purpose of this work to extend the embedding of octonions in Clifford geometric algebra $Cl(3,0)$ of Sect. 3 to a full embedding of the OFT. An essential first step is the question on how to identify the three unit octonions $\mathbf{e}_1, \mathbf{e}_2$, and $\mathbf{e}_4$ with corresponding non-scalar basis elements of $Cl(3,0)$. In [5], page 70, when defining the OFT, it is emphasized that the choice of $\mathbf{e}_1, \mathbf{e}_2$, and $\mathbf{e}_4$, for constructing the transformation kernel is not unique, but other triplets suggested always include $\mathbf{e}_2$, located at the center of the Fano diagram Fig. 2.2 in [5]. Comparing this situation with our Fano diagram Fig. 1, we conveniently choose the three basis blades $\sigma_1, -I, -I\sigma_3$. We observe that

$\sigma_1, -I \in Cl^-(3,0)$ are both odd-, and $-I\sigma_3 \in Cl^+(3,0)$ is even graded, respectively.

We therefore define the embedding in the geometric algebra $Cl(3,0)$ of the OFT of a real signal $f \in L^1(\mathbb{R}^3, \mathbb{R})$ as

$$\mathcal{F}\{f\}(\mathbf{u}) = \int_{\mathbb{R}^3} f(\mathbf{x}) e^{-\sigma_1 2\pi u_1 x_1} \star e^{I 2\pi u_2 x_2} \star e^{I\sigma_3 2\pi u_3 x_3} d^3x. \tag{21}$$

The kernel of the embedded OFT can be expressed in geometric algebra, using multiplication table Table 1, as

$$\begin{aligned}
K &= [e^{-\sigma_1 2\pi u_1 x_1} \star e^{I2\pi u_2 x_2}] \star e^{I\sigma_3 2\pi u_3 x_3} \\
&= [(c_1 - \sigma_1 s_1) \star (c_2 + I s_2)] \star (c_3 + I\sigma_3 s_3) \\
&= c_1c_2c_3 - s_1c_2c_3\sigma_1 + c_1s_2c_3 I + c_1c_2s_3 I\sigma_3 - s_1s_2c_3\sigma_1 \star I - s_1c_2s_3\sigma_1 \star (I\sigma_3) \\
&\quad + c_1s_2s_3 I \star (I\sigma_3) - s_1s_2s_3[\sigma_1 \star I] \star (I\sigma_3) \\
&= c_1c_2c_3 - s_1c_2c_3\sigma_1 + c_1s_2c_3 I + c_1c_2s_3 I\sigma_3 - s_1s_2c_3 I\sigma_1 + s_1c_2s_3\sigma_2 \\
&\quad + c_1s_2s_3\sigma_3 - s_1s_2s_3 I\sigma_2 \\
&= c_1c_3(c_2 + s_2 I) - s_1c_3(c_2 + s_2 I)\sigma_1 + s_1s_3(c_2 - s_2 I)\sigma_2 + c_1s_3(c_2 - s_2 I)I\sigma_3 \\
&= c_3(c_1 - s_1\sigma_1)(c_2 + s_2 I) + s_3(s_1\sigma_2 + c_1\sigma_1\sigma_2)(c_2 - s_2 I) \\
&= c_3(c_1 - s_1\sigma_1)(c_2 + s_2 I) + s_3\sigma_1\sigma_2(c_1 - s_1\sigma_1)(c_2 - s_2 I) \\
&= [c_3(c_2 + s_2 I) + s_3\sigma_1\sigma_2(c_2 - s_2 I)](c_1 - s_1\sigma_1) \\
&= [c_3 e^{I2\pi u_2 x_2} + s_3 I\sigma_3 e^{-I2\pi u_2 x_2}](c_1 - s_1\sigma_1)
\end{aligned} \tag{22}$$

New we observe that to change the sign of any of the three frequency components in the result, GA has very simple involutions

$$\begin{aligned}
K(-u_1, u_2, u_3) &= \sigma_3 K(u_1, u_2, u_3)\sigma_3, & K(u_1, -u_2, u_3) &= \sigma_3 \widehat{K}(u_1, u_2, u_3)\sigma_3, \\
K(u_1, u_2, -u_3) &= \sigma_1 K(u_1, u_2, u_3)\sigma_1, & K(-u_1, -u_2, u_3) &= \widehat{K}(u_1, u_2, u_3), \\
K(-u_1, u_2, -u_3) &= \sigma_2 K(u_1, u_2, u_3)\sigma_2, & K(u_1, -u_2, -u_3) &= \sigma_2 \widehat{K}(u_1, u_2, u_3)\sigma_2, \\
K(-u_1, -u_2, -u_3) &= \sigma_1 \widehat{K}(u_1, u_2, u_3)\sigma_1,
\end{aligned} \tag{23}$$

with grade involution $\widehat{K}$ that changes the sign of all odd grade parts. Note that the frequency sign change only operating in octonion algebra always requires a composition of *four* conjugations (as e.g. in (15)).

5.1 Embedding of Octonion Analytic Signal in Geometric Algebra $Cl(3,0)$

We now ask how the octonion analytic signal, defined in (19), can be embedded in the geometric algebra $Cl(3,0)$ of three-dimensional Euclidean space $\mathbb{R}^3$? Similar to our study of the kernel of the embedding of the OFT, we therefore need to apply the embedding of octonion multiplication in geometric algebra to the convolution factor product that appears in the definition of the octonion analytic signal (19). We again replace $\mathbf{e}_1, \mathbf{e}_2$, and $\mathbf{e}_4$, by the three $Cl(3,0)$ basis blades σ_1, $-I$, and $-I\sigma_3$, and obtain[3]

$$\begin{aligned}
&\{[\delta(x_1) + \sigma_1 \frac{1}{\pi x_1}] \star [\delta(x_2) - I\frac{1}{\pi x_2}]\} \star [\delta(x_3) - I\sigma_3 \frac{1}{\pi x_3}] \\
&= \big[\delta(x_3)\big(\delta(x_2) - I\frac{1}{\pi x_2}\big) - I\sigma_3 \frac{1}{\pi x_3}\big(\delta(x_2) + I\frac{1}{\pi x_2}\big)\big]\big(\delta(x_1) + \sigma_1 \frac{1}{\pi x_1}\big). \qquad (24)
\end{aligned}$$

The following threefold convolution, carried out algebraically in the geometric algebra $Cl(3,0)$, will therefore give the embedding of the octonion analytic signal of (19) in $Cl(3,0)$

$$\begin{aligned}
\psi(x_1,x_2,x_3)_1 = &\big[\delta(x_3)\big(\delta(x_2) - I\frac{1}{\pi x_2}\big) - I\sigma_3 \frac{1}{\pi x_3}\big(\delta(x_2) + I\frac{1}{\pi x_2}\big)\big] \\
&\big(\delta(x_1) + \sigma_1 \frac{1}{\pi x_1}\big) \circledast \circledast \circledast f(x_1,x_2,x_3) \qquad (25) \\
= &f + v_1\sigma_1 - v_2 I - v_3 I\sigma_3 - v_{12} I\sigma_1 + v_{13}\sigma_2 + v_{23}\sigma_3 + vI\sigma_2.
\end{aligned}$$

Furthermore, the seven simple GA involutions of (23) will also analogously yield the embedded version of the octonion analytic signal for the other seven octants, which corresponds to changing one, two or all three signs of σ_1, $-I$, and $-I\sigma_3$, in (25):

$$\begin{aligned}
&\psi(x_1,x_2,x_3)_2 = \sigma_3 \psi(x_1,x_2,x_3)_1 \sigma_3, \quad \psi(x_1,x_2,x_3)_3 = \sigma_3 \widehat{\psi}(x_1,x_2,x_3)_1 \sigma_3, \\
&\psi(x_1,x_2,x_3)_4 = \widehat{\psi}(x_1,x_2,x_3)_1, \quad \psi(x_1,x_2,x_3)_5 = \sigma_1 \psi(x_1,x_2,x_3)_1 \sigma_1, \\
&\psi(x_1,x_2,x_3)_6 = \sigma_2 \psi(x_1,x_2,x_3)_1 \sigma_2, \quad \psi(x_1,x_2,x_3)_7 = \sigma_2 \widehat{\psi}(x_1,x_2,x_3)_1 \sigma_2, \\
&\psi(x_1,x_2,x_3)_8 = \sigma_1 \widehat{\psi}(x_1,x_2,x_3)_1 \sigma_1, \qquad (26)
\end{aligned}$$

where in number ordering of the octants we simply follow Fig. 4.10 and Table 5.4 of [5]. The original scalar signal can always be reconstructed from the eight octant specific signals of (25) and (26), and therefore from the purely positive frequency (in the first octant of the three-dimensional frequency space) signal $\psi(x_1,x_2,x_3)_1$, as

$$f(x_1,x_2,x_3) = \frac{1}{8}\sum_{k=1}^{8} \psi(x)_k, \qquad (27)$$

which is the octant generalization of the reconstruction (18) of a real one-dimensional signal from its complex analytic signal. The single complex conjugation in (18) is replaced by the seven geometric algebra involutions of (26).

6 Polar Representation of Embedded Octonion Analytic Signal

As shown in [16], Theorem 1, there exists an elegant and very compact polar decomposition for complex biquaternions. Due to the isomorphism between com-

[3] Note the close algebraic analogy to the computation in (22), associating c_k and $\delta(x_k)$, as well as s_k and $-1/(\pi x_k)$, for $k = 1,2,3$.

plex biquaternions and the Clifford algebra $Cl(3,0)$, this can be carried over to multivectors in $Cl(3,0)$ as well, see [9], Sect. 4.3, equation (49).

As for notation, unit vectors u (two degrees of freedom (DOF)), unit bivectors i_2 (two DOF), and the central unit pseudoscalar $I = \sigma_{123}$ in $Cl(3,0)$ square to

$$u^2 = +1, \qquad i_2^2 = -1, \qquad I^2 = -1. \tag{28}$$

The even subalgebra of $Cl(3,0)$ is *isomorphic to quaternions* $\mathbb{H}$: $Cl_2(3,0) \cong \mathbb{H}$. That means general multivectors M in $Cl(3,0)$ can always be represented as complex ($I^2 = -1$) (bi)quaternions:

$$M = M_+ + M_- = p + Iq, \tag{29}$$

where p and q are (isomorphic to) quaternions

$$p = M_+ = a_p e^{\alpha_p i_p}, \quad q = I^{-1}M_- = a_q e^{\alpha_q i_q}, \quad a_p, a_q \in \mathbb{R}_0^+, \quad i_p^2 = i_q^2 = -1, \tag{30}$$

with unit bivectors $i_p, i_q \in Cl_2(3,0)$.

The polar decomposition of $M \in Cl(3,0)$ is

$$M = p + Iq = \begin{cases} e^{\alpha_0} e^{\alpha_2 i_2} & \text{for } \; q = 0, \\ I e^{\alpha_0} e^{\alpha_2 i_2} & \text{for } \; p = 0, \\ e^{\alpha_0} e^{\alpha_2 i_2} \frac{1+I\mathbf{f}}{2} & \text{for } \; q = p\mathbf{f}, \\ e^{\alpha_0} e^{\alpha_1 u'} e^{\alpha_2 i_2} e^{\alpha_3 I} & \text{otherwise.} \end{cases} \tag{31}$$

where in line three (compare (26) in [9]) we have the special case that the quotient $p^{-1}q$ results in a unit bivector $\mathbf{f} = p^{-1}q$. The value of $i_2 = i_p$ in lines one (compare (19) in [9]) and three, $i_2 = i_P$ in line four, while in line two we have $i_2 = i_q$. We note that line one is a special case of line four for $\alpha_1 = \alpha_3 = 0$. Line two (compare (19) in [9]) is a special case of line four for $\alpha_1 = 0$ and $\alpha_3 = \pi/2$. So essentially only lines three and four of (31) matter, and we have one special (line three) case with idempotent factor ($\frac{1+I\mathbf{f}}{2}$) and one general case (line four: see Sect. 4.2 of [9] for all computational details) with full exponential factorization. The latter has the necessary eight DOF: four DOF are given by the phase angles α_k, $k = 0, 1, 2, 3$, two DOF by unit vector u' and two by unit bivector i_2.

To better understand how to compute the generic case decomposition of line four of (31), we present the following numerical example (see details in Appendix A).

Example 1.

$$\begin{aligned} M &= 1 + 2\sigma_1 + 3\sigma_2 + 4I\sigma_1 + 5I\sigma_3 + 6I = e^{1.0436}\, e^{1.5574\, u'}\, e^{-0.66405\, i_2}\, e^{1.8304\, I}, \\ u' &= 0.9047\,\sigma_1 - 0.1544\,\sigma_2 + 0.3972\,\sigma_3, \\ i_2 &= 0.2959 I\sigma_3 + 0.6685 I\sigma_2 + 0.6823 I\sigma_1. \end{aligned} \tag{32}$$

We thus propose to use this new polar representation method (31) for the embedded octonion analytic signal (25), is *one way to answer* the open question for the exact polar representation of (19).

Another way to answer this question can be proposed based on analysis of a separable three-dimensional signal $f(x_1, x_2, x_3) = g_1(x_1)g_2(x_2)g_3(x_3)$, $g_k \in L^1(\mathbb{R}^1, \mathbb{R}^1)$, $k = 1, 2, 3$, that leads to a decomposition of the form

$$\psi_1 = a_1 a_2 a_3 \big[\cos(\alpha_2)e^{-\alpha_3 I} - \sin(\alpha_2) I\sigma_3 e^{\alpha_3 I}\big]\big(\cos(\alpha_1) + \sin(\alpha_1)\sigma_1\big), \quad (33)$$

or more general

$$\psi_1 = A\big[\cos(\alpha_2)e^{-\alpha_3 I} + \sin(\alpha_2) i_2 e^{\alpha_3 I}\big]\big(\cos(\alpha_1) + \sin(\alpha_1)u\big), \quad (34)$$

with suitably defined amplitudes $a_k, A \in \mathbb{R}$, angles $\alpha_k \in \mathbb{R}$, $k = 1, 2, 3$, unit vector $u \in \mathbb{R}^3$, and unit bivector $i_2 \in Cl^2(3, 0)$.

Further research has to show which of these two ways may be preferable.

7 Conclusion

We have briefly reviewed octonions and their new minimal embedding in the geometric algebra of three-dimensional space $Cl(3, 0)$. We further reviewed the notion of OFT and octonion analytic signal, embedded both in $Cl(3, 0)$, and finally suggested two interesting possibilities for polar decompositions of the embedded octonion analytic signal. Further research, including concrete applications to non-separable signals, is desirable.

Acknowledgments. I thank God for His help: (Jesus said:) *For everyone who asks receives; the one who seeks finds; and to the one who knocks, the door will be opened* [14]. I thank my colleague S. J. Sangwine for fruitful and inspiring collaboration, as well as the organizers of the workshop ENGAGE 2022, and of CGI 2022.

A Computation of Example 1

The $Cl(3, 0)$ multivector

$$M = 1 + 2\sigma_1 + 3\sigma_2 + 4I\sigma_1 + 5I\sigma_3 + 6I \quad (35)$$

has according to (30)

$$p = 1 + 4I\sigma_1 + 5I\sigma_3, \qquad q = I^{-1}(2\sigma_1 + 3\sigma_2 + 6I) = 6 - 2I\sigma_1 - 3I\sigma_2. \quad (36)$$

A first step is to norm M by division with the square root of $M\overline{M}$.

$$\begin{aligned} M\overline{M} &= (1 + 2\sigma_1 + 3\sigma_2 + 4I\sigma_1 + 5I\sigma_3 + 6I)(1 - 2\sigma_1 - 3\sigma_2 - 4I\sigma_1 - 5I\sigma_3 + 6I) \\ &= 1 - 4 - 9 + 16 + 25 - 36 + I(12 - 16) = -7 - 4I \\ &= \sqrt{65}\,\frac{-7 - 4I}{\sqrt{65}} = e^{2\times 1.0436} e^{2\times 1.8304 I}, \end{aligned} \quad (37)$$

showing that $\alpha_0 = 1.0436$ and $\alpha_3 = 1.8304$. We therefore have

$$\sqrt{M\overline{M}} = e^{1.0436}\, e^{1.8304 I}, \quad (38)$$

and

$$
\begin{aligned}
N = M\sqrt{M\overline{M}}^{-1} &= (p+Iq)e^{-1.0436}e^{-1.8304I} \\
&= 1.9519 + 1.1807\sigma_1 - 0.2712\sigma_2 + 1.7019\sigma_3 \\
&\quad - 0.4520I\sigma_3 - 1.0212I\sigma_2 - 1.0424I\sigma_1 - 0.8828I \\
&= N_+ + II^{-1}N_- = P + IQ.
\end{aligned} \tag{39}
$$

Therefore

$$
\begin{aligned}
P &= 1.9519 - 0.4520I\sigma_3 - 1.0212I\sigma_2 - 1.0424I\sigma_1, \\
Q &= -0.8828 - 1.1807I\sigma_1 + 0.2712I\sigma_2 - 1.7019I\sigma_3.
\end{aligned} \tag{40}
$$

And we represent P as a rotor

$$
P = a_P e^{\alpha_P i_P} = 2.4786\, e^{-0.66405\times(0.2959I\sigma_3+0.6685I\sigma_2+0.6823I\sigma_1)}, \tag{41}
$$

that is

$$
\begin{aligned}
a_P &= \sqrt{P\overline{P}} = 2.4786, \quad \alpha_2 = \alpha_P = -0.66405, \\
i_2 &= i_P = 0.2959I\sigma_3 + 0.6685I\sigma_2 + 0.6823I\sigma_1.
\end{aligned} \tag{42}
$$

We will soon need

$$
a_Q = \sqrt{Q\overline{Q}} = 2.2679. \tag{43}
$$

We finally have

$$
e^{\alpha_1 u'} = NP^{-1} = Na_P^{-1}e^{-\alpha_P i_P} = 1 + 0.8278\sigma_1 - 0.1413\sigma_2 + 0.3634\sigma_3, \tag{44}
$$

with unit vector part

$$
u' = \frac{\langle NP^{-1}\rangle_1}{|\langle NP^{-1}\rangle_1|} = 0.9047\,\sigma_1 - 0.1544\,\sigma_2 + 0.3972\,\sigma_3, \tag{45}
$$

and

$$
\alpha_1 = \operatorname{atanh}\frac{a_Q}{a_P} = \operatorname{atanh}\frac{2.2679}{2.4786} = 1.5574. \tag{46}
$$

In summary the polar decomposition gives

$$
\begin{aligned}
M &= e^{1.0436}\, e^{1.5574\,u'}\, e^{-0.66405\,i_2}\, e^{1.8304\,I}, \\
u' &= 0.9047\,\sigma_1 - 0.1544\,\sigma_2 + 0.3972\,\sigma_3, \\
i_2 &= 0.2959I\sigma_3 + 0.6685I\sigma_2 + 0.6823I\sigma_1.
\end{aligned} \tag{47}
$$

All computations have been verified with The Clifford Multivector Toolbox for Matlab [15].

References

1. Ablamowicz, R.: Computations with Clifford and Grassmann Algebras. Adv. Appl. Clifford Algebras **19**, 499–545 (2009). https://doi.org/10.1007/s00006-009-0182-3
2. Aragon-Camarasa, G., Aragon-Gonzalez, G., Aragon, J. L., Rodriguez-Andrade, M.A.: Clifford Algebra with Mathematica, Preprint, version 2 (2018). https://doi.org/10.48550/arXiv.0810.2412
3. Brackx, F., Hitzer, E., Sangwine, S.J.: History of Quaternion and Clifford-Fourier Transforms. In: Hitzer, E., Sangwine, S.J. (eds.) Quaternion and Clifford Fourier Transforms and Wavelets, Trends in Mathematics (TIM), vol. 27, pp. xi-xxvii. Birkhäuser, Basel (2013)
4. Bülow, T., Sommer, G.: Hypercomplex signals-a novel extension of the analytic signal to the multidimensional case. IEEE Trans. Signal Process. **49**(11), 2844–2852 (2001). https://doi.org/10.1109/78.960432
5. Hahn, S.L., Snopek, K.M.: Complex and Hypercomplex Analytic Signals - Theory and Applications. Artech House, Norwood (MA) (2017)
6. Hitzer, E.: Introduction to Clifford's geometric algebra. SICE J. Control, Meas. Syst. Integr. **51**(4), 338–350 (2012). http://arxiv.org/abs/1306.1660
7. E. Hitzer, S.J. Sangwine (eds.), Quaternion and Clifford Fourier Transforms and Wavelets, Trends in Mathematics 27, Birkhäuser, Basel (2013). https://doi.org/10.1007/978-3-0348-0603-9
8. Hitzer, E.: Quaternion and Clifford Fourier Transforms, 1st edn. Foreword by S.J. Sangwine, Chapman and Hall/CRC, London (2021)
9. Hitzer, E.: On factorization of multivectors in $Cl(3,0)$, $Cl(1,2)$ and $Cl(0,3)$, by exponentials and idempotents. Complex Variables Elliptic Equ. **68**, 1–23 (2021). https://doi.org/10.1080/17476933.2021.2001462
10. Hitzer, E., Lavor, C., Hildenbrand, D.: Current survey of Clifford geometric algebra applications. Math. Meth. Appl. Sci. **2022**, 1–31 (2022). https://doi.org/10.1002/mma.8316
11. Hitzer, E.: Extending Lasenby's embedding of octonions in space-time algebra $Cl(1,3)$, to all three- and four dimensional Clifford geometric algebras $Cl(p,q)$, $n = p + q = 3, 4$. Math. Meth. Appl. Sci., 1–24 (2022). https://doi.org/10.1002/mma.8577
12. Lasenby, A.: Some recent results for $SU(3)$ and octonions within the GA approach to the fundamental forces of nature. Math Meth Appl Sci. (2023). https://doi.org/10.1002/mma.8934
13. Lounesto, P.: Clifford Algebras and Spinors, 2nd ed., London Mathematical Society Lecture Note Series, vol. 286. Cambridge University Press, Cambridge (UK) (2001)
14. Matthew Chapter 7 Verse 8, New International Version of the Bible. https://www.biblegateway.com/. Accessed 01 June 2022
15. Sangwine, S.J., Hitzer, E.: Clifford multivector toolbox (for MATLAB). Adv. Appl. Clifford Algebras **27**(1), 539–558 (2016). https://doi.org/10.1007/s00006-016-0666-x
16. Sangwine, S.J., Hitzer, E.: Polar decomposition of complexified quaternions and octonions. Adv. Appl. Clifford Algebras **30**(2), 1–12 (2020). https://doi.org/10.1007/s00006-020-1048-y
17. Shilhavy, B.: https://vaccineimpact.com/2022/44821-dead-4351483-injured-following-covid-19-vaccines-in-european-database-of-adverse-reactions/. Accessed 02 June 2022

Author Index

E. Hitzer et al. (Eds.): ENGAGE 2022, LNCS 13862, p. 135, 2023.
https://doi.org/10.1007/978-3-031-30923-6

Author Index

Printed in the United States
by Baker & Taylor Publisher Services